DATE DUE

NOV 26 '91	FEB 11 199	JUN 30 '99
MAY 04 '93	APR 08 199	SEP 5 '00
MAY 04 '93	APR 30 199	SEP 15 '01
JUN 24 '9	APR 10 '98	DEC 20 '01
NOV 16 '93	APR 22 '98	APR 05 '02
FEB 25 '94	AUG 26 '98	
APR 05 1994	SEP 08 '98	MAR 11 03
JUN 30 '94	NOV 14 '98	MAY 23 '03
SEP 29 199	JAN 25 99	
MAR 24 '95	MAR 24 '99	NOV 07 03
AUG 01 199	APR 12 '99	JAN 09 04
MAY 30 96	MAY 13 '99	
NOV 29 '96	JUN 99	

GAYLORD   PRINTED IN U.S.A.

Salmon, Jill
    The goatkeeper's guide / Jill Salmon. --
2nd ed. -- North Pomfret, Vt. : David &
Charles, c1981.
    168 p. : ill.

        ISBN 0-7153-8055-9

1. Goats.   I. Title.
                        [A]   [B]   [M]   636.39093

THE GOATKEEPER'S GUIDE

THE GOATKEEPER'S GUIDE

JILL SALMON

DAVID & CHARLES
Newton Abbot   London   North Pomfret (Vt)

British Library Cataloguing in Publication Data

Salmon, Jill
  The goatkeeper's guide. – 2nd ed.
  1. Goats
  I. Title
  636.3′908′3   SF383

  ISBN 0–7153–8055–9

First published 1976
Second impression 1976
Third impression 1978
Second edition 1981

Printed in Great Britain
by Redwood Burn Limited, Trowbridge & Esher
for David & Charles (Publishers) Limited
Brunel House   Newton Abbot   Devon

Published in the United States of America
by David & Charles Inc
North Pomfret   Vermont 05053   USA

CONTENTS

LIST OF ILLUSTRATIONS

Plates

(All illustrations by the author except where stated)

Figures

List of Illustrations

ACKNOWLEDGEMENTS

I should like to express my personal thanks, on behalf of goat-keepers in the South West, to Dr J. B. Tracey, a founder member of the original South Western Counties Goat Society. No problem, whether large or small, was ever too much trouble for him to deal with. His advice to people suffering from allergies and delicate digestions has been of untold value. He has kindly allowed me to condense two pamphlets written by him and distributed by the British Goat Society. An enormous debt of gratitude must go to Dr Tracey and the late Professor Caldwell for starting what has eventually turned into four thriving Affiliated Societies in the South West.

I owe a debt of gratitude too to Hunter Duncan MRCVS, of Rothesay, Scotland, who 'vetted' Chapter 11. My thanks also to Mrs Jill Duncan for her kind Preface, to Miss Helen Hunt for her most helpful Preface and for making it possible for me to see and learn so much about American goats and goatkeepers, and to Mrs Molly Stevens for the photograph of her AN *Redgills Hebe*. Not forgetting my editor, Emma Wood, for her unfailing sense of humour when it mattered most!

Books used for reference were:

Guss, Samuel B., *Management and Diseases of Dairy Goats*, 1977

Hetherington, Lois, *All About Goats*, 1977

Piggott, R. C. in *The Goatkeeper's Guide,* 1st edition, 1976

Wooldridge, W. R., *Farm Animals in Health and Disease*, 1954.

PREFACE

This book has been written at a time when goatkeeping is increasing in popularity, and new herds are being established all over Britain, possibly owing to the ever-spiralling cost of milk, butter and cream, but more probably because some people have become wary of drinking milk from cows which are constantly having to be treated for all forms of mastitis and, until recently, were suspect for transmitting brucellosis.

I have been keeping goats for more than twenty-five years, and have kept varying sizes of herds in the different ways the author describes. She deals systematically and concisely with the basic knowledge which every goatkeeper needs to have, and also describes the several ways of keeping goats in a simple, practical and effective manner. *The Goatkeeper's Guide* makes a notable contribution to the store of goatkeeping knowledge, and should be on the shelf of every goatkeeper.

Rothesay   JILL DUNCAN
Isle of Bute
1976

PREFACE FOR AMERICAN READERS

Reading the manuscript of this book gave me very real pleasure, as I have spent several weeks on the author's farm and have seen most of what she writes about in action. (And I mean *action!*) The amount of produce the author and her partner get out of their few acres of hilly farmland never ceases to amaze me. It is no accident, and all that they have learnt along the way to success you readers can also learn if you add a good measure of elbow-grease and determination.

I have been asked to help USA readers to understand any words or practices that may be unfamiliar. I do not feel a glossary is necessary, but I hope I can help to make your reading more productive and enjoyable.

In the chapter on housing goats, substitute *Fiber Glass* for *Perspex*; asphalt roofing paper or shingles for roofing felt (USA goats like to eat this too!); and keyholes for pop holes. In the chapter on feeding, read corn for maize; alfalfa for lucerne; pellets for nuts; mangel beets for mangolds; and rutabagas for swedes.

If you are going to sell milk or milk products, it should be remembered that there are strict regulations for barns and dairies, which differ from state to state. Before building, you should consult your State's department of agriculture.

Most of the differences in feeds and feeding are caused by the relatively cold and wet growing conditions in England. This makes hay very difficult to dry and harvest compared with conditions over here, and as a result far more green food is fed the year round than in the USA. In England trefoil and alfalfa are fed as green crops, whereas we feed it as high-protein hay. For advice on fodder you should contact your local County Extension Service agent, who will know which are the best crops to raise in your area. It is also a good idea to join a local dairy goat club and seek the advice of established breeders, for in some areas they may be more helpful than the extension agents, who are more familiar with the needs of dairy cattle.

If you plan to show, check with your local club about show rules and how to prepare your goats for the ring. Due to the colder climate, goats are never clipped in the British Isles; instead, they are cleaned, groomed and rugged (blanketed).

If you keep in mind that climate, regulations and medical terms are different, and if you use your imagination or your dictionary when you come across unfamiliar words, you will learn much from this book, which should make the road to success shorter and more enjoyable and the lives of many dairy goats healthier and happier. Far too many goats lead miserable lives, due to ignorance and neglect. In few parts of the world are they as well protected as they are in England, thanks to the Agriculture Act 1968 and to the British Goat Society's *Code of Practice*, which the author was instrumental in drawing up and promoting.

All dairy goats have found a friend in the author of *The Goatkeeper's Guide*.

Washington   HELEN C. HUNT
Connecticut
1976

12

1

BEGINNING GOATKEEPING

Kidding

In these times of 'artificial' foods and flavourings, the goat is becoming more and more popular for supplying an unadulterated source of milk, cream, butter, yoghourt, cheese and meat. Goats' milk is also excellent for rearing babies, lambs, calves, pigs, rabbits and sundry other animals, to say nothing of its beneficial effects on certain digestive disorders and beef allergies. Goat skins can be cured and used, and the manure is excellent for the crops you grow.

That is the rosy side of goatkeeping; now for some of the butts! My partner and I bought our first goats in 1943, and since 1946 we have helped to run various goat societies. During that time we have come into contact with hundreds of people who hopefully started smallholding, but sad to relate very very few survived the rigours of this kind of life. They were either not physically strong enough to stay the course, or they found the daily routine demands of stock keeping more than they bargained for, or their ignorance led them into serious health problems with their goats, which rendered the animals value-less and useless.

By far the most usual reason for wanting to keep a goat is to use it as a lawn mower and weed controller. A goat will never be either, because it is a highly selective feeder and only nibbles the tops off grasses and weeds. A goat seldom grazes grass tight to the ground, as do horses, cattle and sheep. As a clearer of brambles, it will only eat the leaves and the young shoots, leaving the briars to be cleared by man. For people who are not interested in the goat or its produce, who just want something that will clear their ground and who are not prepared to learn to care for the goat properly, keeping a goat is almost always a failure, and a complete disaster for the poor goat.

This is the side of goatkeeping which I dislike. I can never understand why people assume that anybody can farm without knowledge, training or experience. A good stockman may be born with a gift, but it takes years of training and experience to make him really first-rate. Goats, because they are small and easy to handle, affectionate and friendly, are subject to much more abuse than larger animals, and it seems wicked to condemn them to a life in which, through ignorance, they continually change hands and generally go to poorer and poorer homes. I believe that the goat is the most ill-used domestic animal and I hope that this book will go some way towards making a more comfortable life for all goats.

The goatkeeper's first responsibility after his or her family is for the comfort and care of his animals. They must be adequately housed and fed, and fences must be well organised so that the animals do not wander off your property. There is no better way of making enemies than to allow your goats to trespass, and if you do not control them on your own property you will soon be living in a squalid slummy desert! If you tether your goats as being the most satisfactory way for *you* to control them, you will find that they do not milk as well as you were promised, nor will they keep up their yield all the year round. Tethering done by an expert is one thing – they hardly

14

ever do it – but done by a novice it is usually disastrous and the goats hate it.

Anyone contemplating keeping goats must have a love of animals, a certain amount of ground and accommodation, and a desire to use the milk and other milk products which the goat will yield. This book is written mainly for beginners and house-hold goatkeepers, but it also includes an introduction to the fascinating hobby of showing and, for those who wish to develop their goats on a commercial basis, I hope it will give them some guide-lines which will help them along the road to success.

I must stress that goatkeepers past and present have paid far too little attention to dairy work. Many imagine that the goat produces an elixir which all the world is clamouring to buy, no matter how badly it is produced or how awful it tastes! This is quite wrong, and it is as well to remember that one bad sample can lose a good market. People are not all that silly.

There are many things which I have left out of this book, but it has been my aim throughout to try to put newcomers to goatkeeping on a road which I believe is a good one, where the goat is fed properly from the land on which it lives. Call it farming if you like; to my mind too few goatkeepers pay any attention to this aspect of their enterprise. With rising costs and increasing food shortages it is imperative that we should tackle our goats from this more economical angle. There are no short cuts to success with animals or crops and one has to accept that at times the hours are very long and the work hard, but I think it is well worth it.

I hope my book will give you a fairly comprehensive idea about all the things which goatkeeping entails. Obviously you will need to spend money at the start, however simply you intend keeping your goats. You will spend a good deal more time on them during the learning period, but the more you have found out about routine, equipment and other people's methods, and the more preparation you do over housing, fencing and organising your feeding programme *before* you

15

purchase your goats, the greater the chances that your enterprise will be successful and a source of enjoyment to you and your family.

How to start

First of all, get in touch with your local goat club. If you have no idea that there is such a thing, send a stamped addressed envelope to The Secretary, The British Goat Society, Rougham, Bury St Edmunds, Suffolk, asking for a list of affiliated societies. The secretary of the nearest club will be able to supply the names of three reliable people who are willing to show you their goats, and you can then make appointments.

There is nothing more aggravating than to have enthusiastic would-be goatkeepers turning up without warning and wanting to see your goats and pick your brains when you are up to your eyes in work that needs finishing urgently. Study their methods, particularly the crops they are growing and the machinery they use for their cultivations, because it is much cheaper to learn by other people's mistakes. See the way they have adapted their buildings, and try to imagine what it would be like working in pouring rain, snow and howling winds, and then go home and work out your own adaptations with that in mind. There is nothing more pleasing than to be able to shut yourself in the goathouse on a rough winter's night, knowing you have all the food, water and hay that you need without having to potter in and out of that door getting wetter and colder every time you go out.

Find out the cost of hay, straw and concentrates. You will need roughly 5cwt of straw and 15cwt of hay to feed and bed down two goats for a year. A moderate milker, fed with an adequate supply of good greenstuffs, will need about 3lb of concentrates per day, for the whole year. If you intend to grow your own hay and forage crops, you will need about one acre per goat. The fewer goats you keep the easier it is to feed them well on home-grown fodder.

Plate 1 Goats of the author's herd free-ranging
Plate 2 The same herd strip grazing with an electric fence

Plate 3 (Above) British Toggenburg
Plate 4 (centre) An imported pure Toggenburg
Plate 5 (below) Her son, born in this country

If it is possible, visit a herd which is kept for commercial reasons, that is one which produces milk, yoghourt, butter and cheese for sale *all the year round*, because here you will pick up a great deal of useful information.

When planning and costing your venture, do not forget to allow for mistakes, for the times you fail to get your goat successfully mated, for stud fees and for the veterinary surgeon's fees, even if it is only to get your goats vaccinated and your kids dehorned.

For someone wishing really to make a go of commercial goatkeeping, my advice is to get some training first. If you can, do a year on a general farm, then go to a farm institute for a year and specialise in dairying and after that get a job on a first-class goat farm and learn the business from the inside. So many people think they can jump in off the deep end and succeed, and I am afraid that this never works, for you must learn to walk before you can run.

In all of these groups of goatkeepers you will find those who show their animals. It is often a sideline or a hobby, and sometimes a full-time occupation. Successful exhibitors command a high price for their stock both on the home market and for export. But don't imagine that by buying top-flight show goats you will be able to take top show honours and command the best prices for your stock *unless* you have learnt the art of goat-husbandry beforehand, and have then made a name for yourself in the show ring and with milk records.

MANAGEMENT TIPS

The daily routine of feeding and milking, morning and evening is the same for herds kept in all types of housing. Other work will depend on how you propose to keep your goats. It is important, *before* purchasing any animals, to look at your land and decide upon the best method of utilising it. Following are listed the different methods of grazing, and you should preferably make up your mind which is the best for your land before buying any goats.

Zero grazing Goats housed in loose boxes and exercised in yards have all their food carried to them. If you grow all your own crops for cutting, such as grass, lucerne, kale, etc, the system is known as zero grazing. You cut all the food and carry it to the animal. There is obviously a certain amount of labour involved this way, but the advantage is that far less food is wasted so that adequate quantities can be grown on smaller plots of ground.

Paddock grazing If you are going to make small paddocks for your goats to graze, so that each enclosure can be eaten down in a week, you will find that goats do not graze the grass tight to the ground, and you will either have to put in calves or a pony for the second week, or run a mower over the grass as soon as the goats have moved on.

Strip grazing is done with an electric fence. The strip allowed per day is not more than two feet wide, the length depends on the number of goats, and you learn how much they need by the amount left uneaten at the end of the day. Far more grass is consumed by goats this way than by paddock grazing; by only being allocated such a narrow strip the goats do not trample the grass they are meant to eat, and they crop the grass much closer. You can strip graze other crops as well. The ideal method of operation is to put an electric wire behind the goats after seven days' grazing so that they only occupy one piece of ground for a week at the most, which makes for very effective worm control of the pasture.

A very good way of supplying your goats with live food, if you have no ground or time to grow crops, is greengrocers' waste. This can be collected once or twice a week by arrangement with your greengrocer. It is a good idea to feed this in tubs in the yard. Let the goats pick out what they want and put the leavings on the compost heap. One breeder we know used to help herself to the better sticks of celery, lettuce, peas, beans, oranges, apples, etc, and seldom had to purchase any fruit or vegetable for herself!

It is most important to see that stall or yarded goats get

exercise. Taking them for walks is a time-consuming activity if you are busy, but it is also a very relaxing occupation, teaches you so much about the sort of things goats like and gives the goats the opportunity to eat the kind of food you may have omitted from their diet. But, please remember to obtain the owner's permission before taking your goats on a browsing expedition, if the wood or copse does not belong to you.

Official regulations

There are stiff penalties if you fail to notify the Ministry veterinary surgeon, etc, of an outbreak of any of the following: anthrax, foot and mouth disease, cattle plague (rinderpest), and rabies.

All movements of goats from their home farm must be recorded in a book – Movement of Stock Book – kept solely for this purpose. It must be kept up to date. It includes the following: transport to market or slaughter house, or delivery when taking a goat to a new owner, etc; taking a goat for service; transport to and from shows. The police may ask you to produce this book at any time.

2

HOUSING AND CONTROL

There is no point in providing readers with a design for a goat-house, since hardly any buildings are ever erected purely for housing goats. Instead, I propose to give guide lines which can be used whatever type of building is adapted.

From the goat's point of view, light, ventilation and an absence of draughts and dampness are most essential. It should be borne in mind that body heat in any animal is provided by the food it eats. If you keep a goat in a draughty, cold shed she will use more of the costly food you feed her to keep herself warm at the expense of her milk yield. Time, thought and money spent on housing are well rewarded since it is a once-only expense, and you should reap the benefit when your milk supply keeps going during the difficult winter months, without your having to feed vastly increased amounts of concentrates.

Light also has a bearing on milk yields in the same way as it does on laying hens. If you regulate the light to hens you trick them into thinking it is summer and they lay their eggs accordingly. I am not suggesting you do this with your goats, but obviously a dark house will put them off their stroke. Ventilation without draughts is also essential. You will have to sort this one out according to your own circumstances, but if you are going to erect a purpose-built goathouse you need to get the help of the ADAS on this subject.

Pens should be about five feet square. It is not necessary to have them larger, as you use so much more straw, but anything smaller makes the pens difficult for you to work in and prevents the goats from having adequate space to move around. The walls should be 4ft 6in high. I prefer the solid part to be about four feet high and the rest either a strong rail or a rigid weldmesh netting, as this allows the goats to see each other and makes for a much happier atmosphere. It is a good idea to put in a large hay rack so that, in the event of your having to keep the goats housed all day, you can put up enough green food and hay to last them for the whole time between milkings. Provision should be made for a water bucket in each pen.

The doors should, if possible, open outwards. If you use a self-shutting fastener which you can operate from inside or outside the door, do remember that the goat is capable of working that one out for herself. You need also to fix a barrel bolt lower down where it is inaccessible to her! The barrel bolt will also serve as a brace for the gate when she stands with her front feet on the door, which she will always do, and prevent a lot of strain being put on the hinges. An ingenious device is to screw a short length of angle-iron on the gatepost, and a small piece of metal on the gate which runs up on to the angle-iron and makes both a door-stop and a brace. (See Fig 1.)

In designing your accommodation, you should plan where your feed bins will be, where you will store your hay, straw and roots, where you put your water supply and where you are going to 'process' your milk. Also remember that, whatever kind of flooring, you need to 'muck out' regularly, and you should make it possible to use a muck fork without knocking the skin off your fingers every time you turn round. It is helpful if you can have a wheel-barrow in the gangway.

If your space is very limited it is possible to make the partitions between your pens of very heavy-gauge weldmesh. If you screw a stout wooden rail at the top and the bottom you can put large screw eyes on the partitions and the upright posts and have rods which slot through them. (See Fig 2.) In this way the partitions are easily removed for mucking out, or you can make two pens into one for kiddings, rearing several kids, or housing a calf.

Ideally you should be able to reach your goat house without going outside, and once in it you should be able to do all the milking and feeding without opening the outside door.

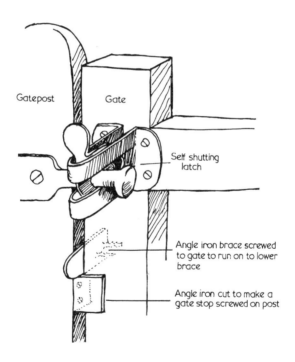

Gatepost   Gate

Self shutting latch

Angle iron brace screwed to gate to run on to lower brace

Angle iron cut to make a gate stop screwed on post

1 A goat-proof gate

24

There are no regulations concerning the layout of goat-houses, milking sheds and dairies, but if you sell or give

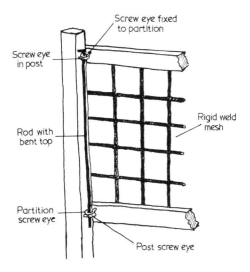

2 An easily-removable partition

away milk and the public analyst is given a sample of it which does not come up to standard, then you will have to have the same facilities – milking parlour and separate dairy with acceptable washable ceilings, walls and floors, running water, etc – as are laid down in ministry regulations for dairy cows, before you will be allowed to sell any of your milk or milk products to the public. Bearing this in mind, it is as well to arrange your accommodation so that your feed and forage store is on one side of the goathouse and on the other side, through a door, is the milking shed, which leads on through another door to your dairy. The idea of keeping the hay and straw away from the milking shed and dairy is to keep dust away.

The best flooring is concrete, draining away from the pens into gutters at the side of the gangway. Without the gutters the effluent seeps all over the passage, making the goathouse smelly and messy. If you are laying your own concrete, a lemonade bottle is splendid for making the gutter! The best bedding is barley straw made into a deep litter. Some people put a layer of peat moss down first with the straw above, but this can be very expensive. We use all the top dryish bedding to put in the bottom after cleaning out and we then put a very generous layer of new straw on top. After that the top layer is shaken daily and any very large heaps of muck are removed and put on the dungheap. If necessary a little new straw is added. We continue like this for four to six weeks and then we muck out, wash the stall and start again. The muck is carefully stacked on a muck heap which, when rotted down, is spread on the ground. The advantage of this method is that the manure produced is excellent for growing crops and is highly saleable, and the goathouse is always sweet smelling – except on mucking out day, and that doesn't last once the new straw is down. If you use oat straw you will find the goathouse gets very smelly. If wheat straw is used, grain left in the ears after threshing can cause digestive upsets, and the goats get very stained because the straw is not so absorbent. If you allow your goats to bed themselves down on hay, not only are you feeding them too much, or feeding them unpalatable hay, but you will have a very smelly lot of goats too! All these forms of bedding make good saleable manure. If, for economy reasons, you use wood shavings or sawdust, you will find that they do not rot in the muck heap, and the manure then upsets the soil structure if used on your land; and if you want to sell it you will find difficulties in getting a customer. Bracken is not advisable because when eaten it can cause internal bleeding and toxic upsets.

Other forms of flooring have been suggested from time to time, but these seem to me to be most unsatisfactory. Some people advocate concrete floors with wooden benches. Provided the goats have small udders they appear quite happy, but the stalls have to be washed out and the benches scrubbed daily. This is a time-wasting chore and quite heavy work, as the benches must be pretty solid if they are to work satisfactorily. The goats get dirty, stained and smelly however much you wash the benches, and you get a very poor yield of manure which has to be moved with a shovel, there being nothing to hold it together.

An even smellier alternative is to put the goats on a slatted floor, which creates the problem of shifting the mess underneath and is much more difficult to wash. Large udders must be at a disadvantage too and both these last forms of 'bedding' must be very cold in winter and covered with flies in the summer.

When siting your goathouse, try to arrange access for vehicles to deliver hay, straw, roots and feeding stuffs and to cart away manure. Ideally the goathouse should be near to water and electricity supplies. Also, and this applies more to the smaller establishments, do not think you can manage with your pens tied up with string and with odd bits of wood and temporary gates. Goats are active climbers and dedicated nibblers. They can slip out of little cracks when you are struggling to open a barrier, and there is nothing worse than having to climb from one pen to another when you are feeding. You waste enough time in one week coping with that kind of set-up to pay for a properly built pen and a properly hung gate.

Some people think stalls are an economical way of housing goats. I don't believe a goat should ever be tied when it is housed, but if they are kept in stalls each goat needs a strong collar with a short chain and a swivel fixed to a running bar. Each time the goats are brought in or let out they have to be individually tied and untied, which is a time-wasting chore. I think that it borders on cruelty, the nature of the goat being what it is, to tether a goat by day and tie it in a stall at night.

One huge goatherd I visited in America had an L-shaped building round two sides of an exercising yard, one end of the building opening onto a courtyard leading to the road. At this end was the loading bay from the dairy. Next to that was the pasteurising plant and then the milking parlour. The rest of the barn was divided lengthwise by hay racks. On the inside the floor was deep-littered, and on the yard side there was a plain concrete floor. Feeding stands for 'haylage' were placed at intervals, and also self-filling baths of warm molassed water. There were great sliding doors which opened onto the yard and there were various bits of apparatus for the goats to climb on in the yard. At milking times the goats were shut out of the deep-litter area, and let in to the milking shed six at a time, where they were fed, milked by machine and recorded, and any sick ones noted and dealt with. They were then let out into the deep-litter area, where they stayed until all the milking was finished. There were two huge silos full of haylage which was automatically chopped and fed into a barrow and then taken to the feeding stands. There were over 300 milkers in this herd. The two owners did the milking (2½hrs night and morning) and a man was employed to sweep the yard and shed and keep the haylage, hay and molasses topped up. About once a month a team often with the Vet was employed to pare hoofs, drench for worms, inoculate when necessary, de-louse and do all the other regular routine jobs. A similar type of loose housing could be adopted for the household goat, where she could have a yard for exercise and either a shelter with a hay and greenfeed rack and water, or a door giving access to her own pen.

Another ingenious system I once saw used a shed with three big loose boxes. These had hay-racks and 'pop' holes (for feeding) on the inside wall, as well as an inside access door. The feed bins, water and hay were all in this area and all the feeding could be done without entering the loose boxes. The other side of the shed had doors opening onto the paddock from all three loose boxes. The first door was very small, the second bigger and the third was big enough for a very large goat to get through, so that at feeding times the goats went into different loose boxes according to their sizes. In this way they could be individually fed without any bullying and without having to sort them out manually.

There are many kinds of building materials you can use for goathouses. If you have to use roofing felt, remember the goats will rip it to ribbons if they can get their teeth into it. A very cheap, warm and not unattractive house can be built with those slabs of wood with bark on them which you can get at sawmills. You need to choose stout ones and they should be trimmed so that they will fit pretty closely together, like planks. Make the framework of your house of decent squared timber. Starting on the inside, nail the slabs horizontally from the floor to the eaves, not forgetting to leave holes for the windows and doors. Then cover the outside with roofing felt and nail the rest of the slabs on the outside, this time vertically. The finished building is waterproof, draught proof and very strong indeed. If you use galvanised iron to roof it, you should work out some kind of insulation. Remember again that if the goats can reach it, they will tear it to bits if they can. Insulation will keep it cool in summer and warm in winter.

If you use corrugated perspex in the roof to give extra light, try to put it on the north side, because it gets very hot when the sun shines directly through in the summer. If you are in a cold spot and require the heat in winter, arrange some sort of easily manipulated blind for use in summer. In any case roof lights which can be covered in hot weather help to keep the fly population down.

Goats are very extravagant with hay. They like to grab a huge mouthful and pull it half out through the bars of the rack and then select the more delectable morsels, dropping the rest on the floor. With present prices of hay this drives the goat-keeper out of his mind. Various methods have been tried to combat this. The best distance between the bars in the rack is 2½in. We tried 2½in weldmesh. We wasted nothing because the goats refused to eat anything through it. Probably they weren't hungry anyway and could hold out until we relented. They just tolerate 2½ × 5in weldmesh because they can get their noses in and waggle a mouthful up and down till it comes through, and then the selection process continues, but they don't waste so much. Another idea I saw was a hay rack boxed in above the first six inches so that the goats could only reach the hay at the bottom of the rack. This idea works well. If you have other stock – calves, bullocks or sheep – which are not as fussy as the goats, a hay-saving rack like the one in Fig 3 is the answer. The snag with this one, designed by the late Mr J. R. Egerton, is that it takes up a fair amount of room.

3 Hay-saving rack designed by the late J. R. Egerton

A goat should be able to eat out of the lowest part of the hay rack with her head at its normal height, but the top of the rack is best out of her reach even when she is standing on her hind legs, or she may pull all the hay out of the top on to the bedding and, having stood on it, refuse to eat it. Many goatkeepers fit a lid on to the top of the hay rack to stop this, but it saves time if the goats can't reach it. But you must make it as easy as possible to put hay in the rack without having to go into the pen or open the door. Ideally, you should plan a feeding passage down the middle of the goathouse to which the goats have no access at all. If the racks are made of weldmesh you can bend the tops over about six inches, which nicely stops pilfering.

Fittings for water and concentrates are always a problem. Pop holes through which the goats poke their heads to get to buckets outside their pens are the most satisfactory. Goats will never eat food or drink water which they have fouled. They will starve to death rather than do this. Having the buckets outside the pens makes collection and feeding and watering easy and labour-saving. I once saw automatic drinkers installed in goat pens, and the goats were all suffering from dehydration because the goatkeeper had never cleaned the bowls out. This would be an everlasting job since the goats would be forever standing their dirty feet on the edge of the bowl, if not in the middle.

Mangers are also unsatisfactory for goats as they are difficult to keep clean and most goats like either to sit in them or stand on them.

It is no wonder the idea had got around that the devil has cloven hooves. I think the person who started that one must have been trying to domesticate a goat. If the truth was known, I think goats must have been the inventors of trades unions, for they definitely have one and it has the most intractable rules!

Hay nets are not satisfactory either, because the goats get their feet caught in them and the kids get hung in them. They also are tiresome and time-consuming to fill. We use ours for shows, but not at any other time.

Controlling goats

A goat is an agile, active animal and can jump onto quite high walls and low roofs. Goats seem to take a delight in dancing on tin roofs, slithering down thatched roofs and pulling out the thatch. They can unfasten most gates if they can get their mouths near enough to the catches and they take flying leaps over wire fences Grand National style. Fearful problems crop up when goats try unsuccessfully to leap over barbed wire fences and get their udders caught. As for a nice thick hedge, even thorn presents them with no problems at all; they just eat a hole through it! They have methods of dealing with chain-link, pig netting, sheep wire and ordinary wire netting. The latter, if well erected and five to six feet high, provides them with a delightful way of grooming themselves. They hurl themselves at it sideways and, pushing with all four feet on the ground, they rub one side with great force against the netting until they reach a post. They then turn round and repeat the performance the other way. The result is that the netting gets more and more of a bulge and eventually splits, and then the goat joyfully squeezes through the hole. If you use the heavier types of netting they climb up it, and after a period of such abuse, the netting sags and whoops . . . over she goes. That has effectively dealt with most types of barrier, other than a six-foot-high brick wall, in good order, which would be quite satisfactory, but somewhat inflexible and very costly!

To be serious, there are effective ways of fencing goats in. Pig wire and chain link should never be used for goats without a very strong straining wire or rail at the top, and the bottom needs pegging down too. An expensive, permanent type of goat-proof fence would be a post and three-rail fence with pig netting, chainlink or sheep wire. The bottom rail should be 6in from the ground and the middle one halfway up the goat's flank, the top one 4ft 6in high. The netting should be firmly fixed on the side away from the goats. The top rail prevents the netting from sagging should the goats climb on it; the middle rail stops them rubbing against it and provides them with something to stand on, and the bottom rail serves as an anchor for the netting and prevents them from squeezing under it.

Electric fencing is excellent for goats. It is highly mobile and adaptable. Sometimes when your fence is free-standing one or two enterprising goats find that if they take a speedy dash under the bottom wire it does not hurt. When this happens, put a second wire about two feet away from the first at the same height as the lower wire. Similarly, if they start jumping over the wire, a second strand, parallel with the top wire and also about two feet away will stop that. If you have a hedge, wire-netting fence or similar barrier which is not goat-proof, a single-strand electric fence about two feet away will stop all attacks. They will never charge an electric fence if there is any sort of barrier on the other side.

It is not nearly so costly to erect a single-strand electric wire along a permanent fence, particularly if it has wooden posts. A cross-piece of wood 3ft long, nailed in the middle at about 20 to 24in, with an insulator at either end, will provide you with a permanent place to slot in your wire as required (Fig 4). You can use wood or stout wire for the arms, and either insulators or 2in lengths of ¼in plastic water-pipe, through which, when fixed to the end of the arms, the electric wire can be threaded.

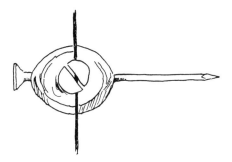

4 Easily-threaded and useful insulator for wooden posts; supplied with nail

I have never made up my mind whether permanent electric fences should be erected, because they need a great deal of attention. You must always be sure that there is no vegetation touching it, otherwise the current is earthed and the fence useless. A very good electrified netting called *Flexinet* is on the market. This is light and highly mobile and fifty yards can be easily carried under one arm. If you run the tractor over the place you wish to put your fence the herbage is flat enough to 'plant' it without further trouble. Failing this the grass should be cut in a strip about a foot wide. The effective life of this netting and also of plastic electric-fencing wire is about three years. The advantage of *Flexinet* is that the stakes are woven into the wire. The disadvantage is that the goats sometimes leap over it. I think this trouble is probably worse with British Alpines and British Toggenburgs and possibly with similar cross-bred types, but this can be cured with a single wire two feet away from it. Another advantage of *Flexinet* is that even when it ceases to give a shock you can still use it as a barrier behind a single strand of electrified wire and it will last for years.

All animals have to be trained to the electric fence, otherwise it is useless. To do this you erect it across the paddock, keeping the goats in a very small piece of ground. This will cause them to investigate. One by one they will examine the wire and get a shock. Stay with them until all have been shocked at least three times, or until they shy away from it and will not go near for further investigation. They have then learnt their lesson. Never try to move the wire if you are strip grazing when the goats are about. They discover it is not shocking and then you have problems.

A great advantage with goats or any animals trained to the electric fence is that if you have to drive them through the garden or similar delectable eating places, a single strand on either side of the path will guide them past temptation, provided you are there to help. The wire does not need to be electrified or fixed to proper stakes, unless you happen to leave them there unattended.

Tethering as a form of control, if well done, is a very time-consuming occupation. To start with, goats don't like high winds or rain. Neither do they appreciate excessive heat or flies, so that you are perpetually on the watch for the weather. Also a goat will not eat herbage which has been trampled down, so that to keep a goat eating all day you need to move the tether about two feet every two hours. Some people think a very long tether gets over this, but it merely allows the goat a

Plate 6 *(Above)* Pure Saanen
Plate 7 *(below right)* British Saanen
Plate 8 *(below left)* Rear view showing tail bones ridging up

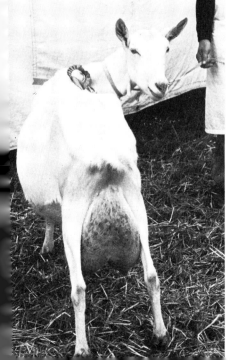

Plate 9 British Alpine
Plate 10 Anglo Nubian

larger area to trample and bruise. A running tether is not advisable either, because mostly this is fixed at either end with the goat attached to a shorter chain which runs freely along the running wire. Not a mobile contraption at the best of times and the goat soon renders the pasture unpalatable.

Goats thoroughly enjoy being taken for walks, browsing from the hedges or in woods, if this is possible. If you ever want to collect them you just pull down a leafy branch and they will all come. They should also follow you if you call them.

Some people have the right to graze their stock on commons. These rights are usually mentioned in the deeds of your property. But, unless specifically mentioned, goats are not commonable animals. If there is a local commoners' association, goat owners who have common rights should discuss the problem with the association and reach a decision that way. Pigs are not commonable either and I think both pigs and goats had in the past to be herded on commons if there were any specific rights to graze.

The law about fencing animals is quite clear. Generally speaking, it is the responsibility of the owner of the stock to keep his animals from straying onto other people's land. One exception to this is that if a landowner sells a small plot in a field he usually gets written into the deeds that the buyer has the responsibility of keeping the farmer's stock off his plot. The other exception concerns common land. It is the responsibility of all landowners whose property borders on common land to maintain stock-proof fencing. If common animals break into Mr A's land and then from Mr A's to Mr B's land, it is Mr A who is held responsible and would have to pay Mr B for any damages he was entitled to claim.

Most insurance companies who deal with farming and the land will issue a policy covering straying stock.

3

WHAT BREED?

Having completed your inquiries, you will be wondering what breed you should keep. This very much depends on your own circumstances. Are you going to keep your goats in conjunction with a self-sufficiency holding? Have you only a very small plot of ground so that your goats will have to be housed and fed by hand? Do you want to keep goats for showing? Are you hoping to build up a commercial herd? The answers to all these questions depend on the amount of land you have and the time you will be able to spend on your animals and cultivations.

Starting with the household goat kept in conjunction with a large garden, or the back-yard goat, both of which will be kept in a shed with an exercising yard, I personally think the more placid, less active Saanens or British Saanens are the answer here. They seem not to mind being restricted. They are white, the pure Saanens being somewhat smaller than the British Saanens, and both types milk well. If you have no intention of showing your goats, cross-bred white goats are quite suitable provided they are from good milking strains. Many of the good-looking non-pedigree goats about are descendants of the vast number of pedigree stud goats which have been used for service over the whole country for the past fifty years or so.

For a larger holding, where you will be able to graze your goats and grow a certain amount of fodder for them, either the Saanen or British Saanen or the Toggenburg or British Toggenburg goats are suitable. All these breeds seem to graze well and are not as troubled by flies as other breeds. This then becomes a question of personal choice.

Anglo-Nubian goats usually have much shorter lactations than the other breeds and most of them need mating every autumn, which means they go dry for a period during the winter. On the other hand their milk contains more butterfat than that of the other breeds, so that if you needed to increase the butterfat content of the bulk of your milk, the addition of Anglo-Nubian milk would help considerably.

I think British Alpines (the goats we keep here) are the most difficult of the lot; but although active, wilful, skittish and highly intelligent – as are most goats – we find them so attractive to look at! They are more bothered by flies than the other breeds and seem to object to too much sunshine. The greatest difficulty is breeding them true to type.

THE DIFFERENT BREEDS OF GOATS IN BRITAIN

Toggenburg (Fig 5a) A small brown and white goat, often with a longish silky coat, short upright ears and a 'dish' nose, with white facial stripes, white legs and a white patch on either side of the tail. Can be horned or hornless. This is a pure-bred goat, which means that only those kids produced by mating pure-bred parents may be registered in this section of the herd book. This breed is descended from imported goats from Switzerland.

British Toggenburg (Fig 5b) A much larger animal than the Toggenburg which was evolved by crossing the Toggenburg with other goats in this country. (Not with Anglo-Nubians.) It

(a) PURE TOGGENBURG

(b) BRITISH TOGGENBURG

(c) PURE SAANEN

(d) PURE SAANEN

(e) BRITISH SAANEN

(f) BRITISH ALPINE

(g) ANGLO NUBIAN

5 Typical heads of the different breeds

has the same markings but its coat is nearly always short. The colour is often darker than the pure Toggenburg and the nose is much straighter. It is possible to breed up from unregistered goats into this section of the *Herd Book* but it takes at least six generations.

Saanen (Figs 5c & d) Another pure breed which was imported from Switzerland. A white goat with a short silky coat, horned or hornless, with a dished nose.

British Saanen (Fig 5e) This was a breed evolved in this country as a result of crossing the Saanen with other white goats. It is a larger animal with longer legs and often a much straighter nose.

British Alpine (Fig 5f) A large black and white glossy-coated animal with the same 'Swiss' markings as the Toggenburgs and British Toggenburgs. This breed has been evolved in Britain from goats which appeared from time to time when the previous breeds were crossed. It is difficult to breed true because there is no 'parent' stock to use.

Anglo-Nubian (Fig 5g) Another breed evolved in this country by crossing native or cross-bred goats with Nubian goats imported from the Middle East. These have very long lop ears, a high 'Roman' nose and very straight legs. They carry their heads much higher than the other breeds and their hind legs often seem to be longer than their front legs, giving the impression of a dip in the back. Sometimes Anglo-Nubians have 'undershot' jaws and 'wry' tails but neither of these are considered faults since they were possessed by the original imports. They can be any colour at all although the Swiss facial stripes are undesirable. It is interesting that, although the last import of this breed was in 1904 the type remains dominant, and if crossed with any of the other breeds with a view to improving butterfats, Anglo-Nubians always transmit their characteristic ears and noses, which are very difficult to breed out.

Golden Guernsey This is the latest breed to arrive in this country. The goats are golden or honey coloured with silky coats, fine bones, a slightly dished face, and ears set like a bonnet with a slight curl at the tip. Breeders believe these traits came from the Syrian goat which was interbred with the 'native' stock on Guernsey hundreds of years ago. The breed very nearly became extinct and its numbers dropped to less than sixty at one point. There are now over two hundred goats on the mainland, and breeders are trying to achieve better conformation and higher milk yields, whilst being greatly restricted by a scarcity of unrelated bloodlines.

Buying a goat

So many people ask what is the right price to pay for a goat, and since prices vary from year to year it is impossible to give a specific answer. At the lowest end of the price list is the small, nondescript goat which only milks for a couple of months or so, and at the top a beautiful, long-lactating animal which has earned all the decorations she can get. To my way of thinking an animal like this could come from any section of the *Herd Book*, ranging from the Identification Register to a Breed Section, and I cannot for the life of me understand why an animal such as this should be worth less if it is registered in one of the lower grades. In all probability its breeding could be as good as that of one in a Breed Section, only someone along the line has forgotten to register one of its ancestors.

It is simple enough to work out what it has cost to rear an animal to the age at which it is being sold, and you should be able to judge the price asked from that. I do think that, at the time of writing, there are some quite idiotic prices being charged for pretty poor animals, and this does no good to either the sellers or the goat industry. If we are going to improve goats we simply must get back to destroying our culls instead of selling them. I think that the general standard of present-day show goats is a good deal lower than it was fifteen years ago, and I am sure that this is due to lack of proper culling on the part of breeders.

When buying a milker the following points should be noted and they apply to goats of all breeds.

She should look alert and healthy. The 'bloom' on her coat and the brightness of her eye should tell you this. She should look feminine, not coarse and, when you handle her coat, it should be soft, loose and supple. She should stand firmly on her feet, which should be inspected to make sure they have been trimmed properly, and that she is not rocking back on her heels. It is most important that a goat should be sound in her feet and legs as she has to carry considerable weights on them when heavily in kid.

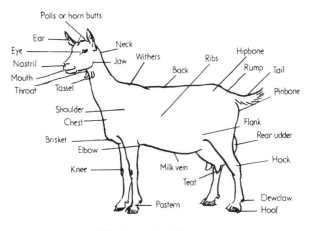

6 Points of a dairy goat

Fig 6 shows the outline of a good Swiss type goat. Note the pretty head with ears carried forward, and large, bright eyes. With goats that have been de-horned, watch for horney growths. Small scurs that are loose do not matter but large deformed horns are a continual source of worry. Never buy a horned goat as they are apt to damage other goats. The neck should be long and slender, and if there are tassels these should be evenly hung. The withers should be fine. Any thickness in

43

the shoulders, or cavity under the shoulder blade is undesirable. The back should be straight, and the slope from the hips to the tail (rump) should be slight. The steeper the drop the worse the fault. A good depth of body is necessary. The udder should have a good 'front' to it and should be joined to the body where the arrow points (on page 45). So often there is a cavity here which is undesirable (Fig 9). When milked out the udder should be soft to handle and free from lumps.

7 Rear view showing the ideal 'wedge' shape

8 Rear view showing good, straight hocks and a well-hung udder

It is most important that the buyer should milk the goat before buying, as a slow milker is a great nuisance and time-waster. The milk should also be tasted.

If a goat is unregistered, it is often possible to tell its age by looking at its front teeth (Fig 10). Up to about one year goats have eight small milk teeth in the bottom jaw. Goats, sheep and cows never grow front teeth in the top jaw. They have a pad which is used in conjunction with the tongue to grip the grass while the teeth in the lower jaw 'cut' the grass. At about a

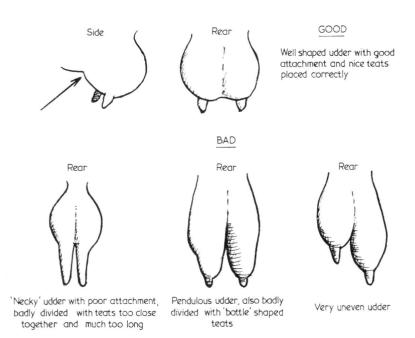

Side   Rear   <u>GOOD</u>

Well shaped udder with good attachment and nice teats placed correctly

<u>BAD</u>

Rear   Rear   Rear

'Necky' udder with poor attachment, badly divided with teats too close together and much too long

Pendulous udder, also badly divided with 'bottle' shaped teats

Very uneven udder

9   Udder shapes

year the centre pair of teeth drop out and are replaced by two much larger permanent incisors. At two and a half years they have four permanent teeth and usually between three and four years they have what is termed a full mouth. After this, age will wear down the teeth, and when they become loose or missing an animal has finished its useful life as a grazer or browser. Obviously these front teeth are not so important for a stall-fed goat because she does not have to 'cut' her own herbage.

A registered goat, particularly if she is earmarked, is no problem. Her birth date is on her registration card, and often her ear number, and this should be compared with the mark in the goat's ear.

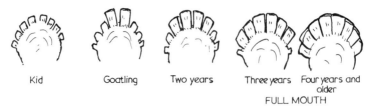

Kid	Goatling	Two years	Three years	Four years and older

FULL MOUTH

10 How to age a goat by its teeth

If you are buying a kid (up to twelve months) or a goatling (up to two years), much of this advice applies, but since it is most unlikely that the animal will have an udder, you should have a good look at the undeveloped teats. They should be placed well apart and there should only be two of them. You should also look carefully at the teeth. If a goatling has four teeth and you are buying her between November and February, ask the seller if and when she has been mated. After three months of the gestation period she should be showing some signs of pregnancy. Her abdomen should give the appearance of being occupied, even if it does not bulge, and the vulva and the skin around it should appear more puffy than that of a goat which is not in kid. If she has been mated and you think she is *not* in kid (this is not an easy thing to decide even for an experienced goatkeeper) you should treat her with caution, or get someone else to look at her. Any goatling (or animal with four permanent teeth or more) which has never kidded should be regarded with great suspicion, because in all probability it is a non-breeder, for one reason or another.

Kids should be examined in the same way, and also the vulva should be looked at to determine whether there is a swelling at the bottom. This also indicates a non-breeder. (See Chapter 9.)

Registered animals are issued with cards by the British Goat Society. On one side are recorded the name of the goat, its birth date, its registration number, its earmark if any, its sire and dam and the name of its breeder. If the breeder has sold the goat you should also find the name and address of the person to whom it was sold. The last name on this side must be the name of the person who is offering the goat to you. If this is not the case then you must insist that the seller gets the omission rectified before you conclude the deal. There is the possibility that the seller may not be able to do this, which will mean that you are being offered an animal which is not registerable in its proper section of the Herd Book, and that should materially affect the purchase price.

My personal opinion is that those wanting to buy a registered goat are well advised to send for a transfer form from the British Goat Society before making the purchase, and to take it along with them. They will then be able to see if the card is in order, fill in the transfer form and get it signed by the seller, and they will then be able to post it with the fee knowing that that part of the deal is all in order. The registration card has to be sent with this form and the whole lot will be sent back to the buyer. The seller can refuse to sign until the goat has been paid for, which is only fair.

It may help here to explain the difference between the words 'pedigree' and 'registered'. Pedigree means of known parentage, and any little scruffy goat can be called pedigree if the sire and dam are known to the seller. Registered means that the goat has been registered as described above, and that it should have a registration card with all the details recorded on it.

4

FEEDING

'Is that really me?' (*Bob Martin*)

Bulk feeds for goats

Grass and hay So many people think that all grass is equally good. This is not so. Yorkshire fog, for example, is only palatable in the early spring growing period, but long before it comes to the flowering stage the food value has fallen and its leaves have become hairy and unpalatable. Cocksfoot is another grass which gets rank and coarse very quickly and it gets tufty and unmanageable in fields grazed by goats.

If you wish to make the most of your grass and intend to save your own hay you should think about having the field resown if necessary. You can get all the advice and help you need about this from the local ADAS service. They will advise you about mixtures for your area and how you should treat the grass as a crop. Your nearest agricultural contractor will quote prices for ploughing, harrowing, muck spreading, cutting, baling and carting hay and so on. If grass is well managed you can get upwards of 100 bales off a one-acre plot, so that with present-day hay prices it is well worth doing and will more than cover the contractor's costs.

When grazing goats you should see that they never stay more than seven days on the same plot. Apart from the fact that the pasture will be getting fouled and be no longer appetising, the worm eggs deposited in the goats' dung will begin to hatch and climb up the grass blades when the grass is damp. These are eaten by the goats, who become reinfested, and so the cycle starts all over again. By strip grazing with the electric fence not only can you prevent this, you can also ration the amount of grass the goats have daily so that they eat it much closer to the ground, thus using more of the pasture for feeding and so letting the sun and air get right down to the ground making the atmosphere too dry for worms to thrive.

We have found that when the grass is growing very fast in May it is often necessary to cut it as soon as the goats have come off it (after seven days). If you have other stock (horses, calves or bullocks) these will do the cutting for you. Sheep do not provide you with a worm control since they suffer from the same worms as goats.

In the spring the grass needs harrowing (which is a form of raking). If you are shutting a field down for hay it should be harrowed prior to this. When to stop grazing is a matter for speculation, but a general guide is that if there has been an early spring and the grass is growing very fast in late March and early April, it is soon enough to shut down towards the end of April. On the other hand if there has been a cold spring and the grass is growing slowly, the earlier in April you stop grazing the better.

Grass crops need feeding. If you make your muck heaps in rotation you should have a well rotted heap ready to spread on the pasture in early winter. You can either cart out the muck and dump it in heaps all over the field for spreading with a fork or you can get a contractor in with a muck spreader to do the job for you. When spread on the grass it should be chain harrowed well as soon as possible after there has been a prolonged shower of rain. If you use artificial fertilisers, the fields can be dressed straight after the hay has been carried and again in February.

The crop is ready to cut when the grass has come to a head and just before the pollen appears. It has the highest food value at this stage. The object of making hay is to conserve this food value, and to do this you have to stop the crop from 'breathing' as soon as possible. The hay should be cut when all dew has gone and then be turned or shaken up at once. In this way the pores on the leaves are sealed and breathing stops. If the swaths are left unturned the grass under the swath does not wither and breathing continues. This converts the proteins into carbohydrates, which spoils the feeding value of the hay, particularly for milkers.

Hay should be turned again the next day when the dew has dried, fluffing up the grass on to the dry wind rows between the swaths leaving a bare path for the next turning. It should be turned again in late afternoon. If the hay has dried well by the following morning it should be rustling, and at this stage two rows can be made into one. All being well weatherwise you should be able to decide if it will be fit to bale the next day and you can make your arrangements accordingly. A good way of testing if it is fit to stack or to bale is to take some strands of hay and twist them very tightly together as if you were making rope. If any moisture is squeezed out the hay is not ready. If you let the hay get too dry it becomes unpalatable. Knowing just when to bring it in is an art which is very hardly learnt and which after a lifetime of experience still causes much head-scratching and heart-searching. After the contractor has done your baling he will tell you if the bales can be carted straight from the baler to the barn or if they should stand in groups in the field for further drying.

Tripod haymaking is ideal for goatkeepers who want to make small quantities of hay, because the whole process of haymaking is done on the tripod and when it is ready to stack in the shed there is no chance of it heating up or going mouldy. It is also a more satisfactory way of curing hay in uncertain weather. You need three poles about seven feet long and three

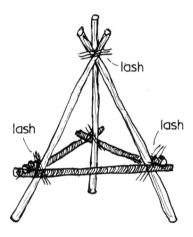

11 Curing hay on a tripod

poles about four feet long. Lash the three long poles near one end as shown in illustration (Fig 11). Lash the three short ones about a foot and a half from the ground. Stuff three large paper sacks with straw and place them under the three cross-poles. Cut the hay when the grass is dry and turn at once. Turn again the same day. Turn twice more the following day, and if the grass has almost begun to rustle it is ready to put on the tripod. Load the grass very carefully along the cross-poles. The sacks will prevent the grass slipping off. Load round and round the tripod so that the height grows evenly all round. Be careful not to fill in the centre, which is the area shaded in Fig 12, as this should be hollow. The sides should be upright and not sloping in towards the top, and the finished height will be as high as you can pitch it up successfully. You must take fairly small forkfuls when you are building the tripod.

Trim down the sides with the pitchfork and make sure you have a well rounded top. Remove the three sacks. This should leave a gap between the ground and the hay and allow air to circulate under and inside the tripod. After two days the hay

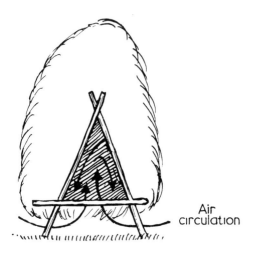

Air circulation

12 Curing hay on a tripod

will have settled, and if the top looks weather-proof and needs no further attention, throw baler twine over the top and secure it to each leg so that in the event of a gale the top stays on. If the tripod blows over get it upright as soon as you can.

Even in good weather tripods make very much better hay than that which is bleached on the ground. This is also an excellent method of drying small quantities since you leave the grass on the tripod until it is properly cured. You can make up to two bales of hay on one tripod.

Hay is the most important bulk feed for goats and it should be the best you can make or buy. If you give them second-class hay they pull it out of the racks and use it for bedding. When buying hay it is a good plan, if you get the chance, to buy one bale and let the goats try it. If they won't eat it, or pull it out all over their pens, don't buy it!

Hay which is carted in before it is properly dried heats up, and can ignite spontaneously. Heating also turns the hay

Plate 11 French Alpine, the Alpine of America. Its head markings are almost
the reverse of the British Alpine
Plate 12 La Mancha, another US breed, hailing originally from Mexico

Plate 13 A good way to hold a kid's head when teaching it to suckle from a bottle

Plate 14 Hold the teat so that, if the kid pulls it out of the bottle, the feeder can prevent it from being swallowed

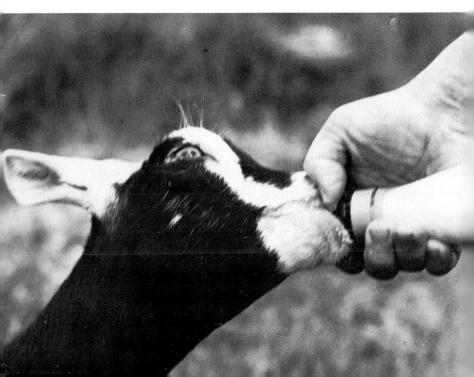

brown, gives it a delicious smell, and raises the sugar content. Such hay is alright for fattening bullocks or lambs, but it is no use for milking animals, they grow fat and don't milk well. Hay which is baled too soon and is too wet sometimes grows a white mould inside the bales if left out to dry in the field. This also happens with small stacks of hay which are too small to heat up. Never buy mouldy hay or straw. I cannot repeat this too often. It makes the animals cough. It can cause breeding troubles and in human beings it causes sinus and bronchial troubles and at worst can give you farmer's lung, which is a very serious complaint indeed.

If you have stacked your hay in a rick and it starts to get hot, (heat up) watch it carefully. If the heating continues you should try to plunge your hand down through the hottest part. If after two or three weeks it is getting increasingly hot, you need to bore the rick. You do this with a bar or hay knife and make a hole big enough to get your arm in, no more, and go down as far as you can. You will find steam comes out for a few days and then the heat subsides.

Green crops   There are many green crops which are excellent to grow for both household or commercial goats. These are cut and fed straight to the goat. Lucerne should be sown in April on a really well cultivated weedless plot. If a few lettuce seeds are mixed with the (inoculated) lucerne seeds, these will come up much more quickly than the lucerne and allow you to hoe between the rows. They can be pulled out as soon as the lucerne seedlings are established. You should get three or four cuttings from this crop during the summer.

Many goatkeepers swear by comfrey, but I must confess that none of our goats have ever eaten it. A permanent bed can be established from cuttings, but it would be as well to try feeding it wilted to your goats before growing it. I understand that once planted comfrey becomes a major pest as a WEED.

Oats and vetches are a splendid green feed, ready in early spring. These are sown in late September or early February, and in a wet summer will give you three good cuttings.

Trifolium incarnatum is another spring green feed. This must be sown in August. It usually yields one crop, but should not be sown in with other grasses as it is a large plant which smothers its neighbours.

Kale is the real standby winter green feed. It should be sown in well worked and well manured ground in April and May. The crop needs the same space as your household greens. If you are going to strip-graze the kale, sow it thinly and keep it weeded between the rows. If you are going to cut it for feeding in racks, it is better to single the crop allowing about a foot between each plant, which will grow much bigger and need less handling when you are cutting.

Mangolds   Most goats like these either chopped or whole. We have found it much cheaper to buy them than to grow them. They should be sown in late April on well manured ground and you must keep the hoe going through them until the end of July if you are to get a good crop. You should allow eighteen inches between rows, and twelve inches between plants.

Parsnips, swedes and carrots are all eaten by goats and, if they don't like them, you can always feed them to your family!

Other green foods from the garden are pea shells, and pea haulms. Some goats will eat broad bean stems and pods, but not many. Some will eat bolting lettuce. All like spinach gone to seed, bolted brussels sprouts before the flowers open, and all trimmings from garden greens. Runner beans and French beans which have gone stringy are also acceptable as are the haulms, potato peelings dried in the oven, apples and pears, and peelings from most fruits. Some goats like carrot tops, but never feed the tops of mangolds, beetroots or fodder beet.

A great many goatkeepers spend endless hours scouring the hedges in the district for browsings. Unless you have specific permission from the owner of the hedge or wood to do this, it is quite illegal, and is a practice which gives goats and their owners a very bad name. Hedges are grown for shelter, to keep in stock or as boundary. Indiscriminate pruning by man or goats will seriously damage them. Also, you cannot possibly gather sufficient browsings to provide enough bulk green feed to fit in with the following method of feeding concentrates.

Feeding concentrates

Articles on feeding concentrates are a bit of a mistake because no one recipe fits the many different ways people manage their goats or the great variations of soil and climate. There are general guide lines, however, best of which is to have a real understanding of the digestive process.

When a ruminant eats, the food reaches the rumen where bacteria use some of its energy, waste some as methane gas and leave the rest as volatile fatty acids which are absorbed into the bloodstream as the main source of energy. They also break down most proteins, reducing them to ammonia of no value to the goat but enabling other bacteria to build up their own body proteins which, when bacteria die, get washed into the goat's intestines and are digested by enzymes to form amino acids. But a high yielding goat cannot get enough amino acids this way and needs proteins not attacked by bacteria but digested in the gut. The best sources are brewers' grains, dried grass, lucerne and soya bean meal. Most digestive upsets are caused by eating a strange food which cannot be processed in the rumen since the right type of bacteria is missing. Thus cows scour (excessive diarrhoea) when suddenly turned out to grass in the spring when previously fed on hay, silage, etc. Sometimes medicines given by mouth destroy the 'flora' in the rumen and this can be corrected by giving brewers' yeast, Cytacon, molasses, and cider vinegar. The quickest way to deflate a 'blown' goat is to give a large dose of brewers' yeast.

Bearing all this in mind, it follows that *any* change in the diet of your goat must be done gradually so that the flora in the rumen can be built up to deal with the change. If you want to turn your goats out onto spring pasture after they have been indoors all winter, ten minutes is long enough for the first day.

Each day they can stay out a little longer, but the changeover should take at least three weeks. This rule applies to any change.

In all methods of management goats should be fed hay right through the year. In summer one feed at night is sufficient.

We have never worked out the actual amount of concentrates, as do most people, on the basis of so much for maintenance and so much for each pound of milk given, as we think that this leads to gross overfeeding of concentrates and, at present prices, this is *very* expensive. We work our feeding programme on a basic ration concentrates which never varies. Individuals may vary, of course, according to size and other factors and we arrive at the basic amount after the goat has kidded. She has no concentrates on the first day and is then given a little crushed oats. By the end of a week she is having a small amount of a mixture of all varieties. We gradually increase this until she leaves some in her bucket when we reduce the total by a handful, leaving us with her normal ration.

By adopting this method you always get all the concentrates cleared up in a few minutes and if this doesn't happen, then you know that there is something wrong such as worms or that the goat is coming into season. When a goat is drying off prior to kidding we still give her the same ration because, as the milk gets less, so the kids get larger.

By using hay and concentrates as a basic ration we can use home-grown feeds for boosting the milk. In the late spring and summer the goats are strip grazed on ley mixtures; they have a certain amount of time in the woods and every evening they have a very large feed of either cut ley mixtures, oats and vetches, white horse tooth maize or anything else we have grown for them. In the autumn we feed kale in the place of these foods and, when the pasture finally ceases to provide them with a useful feed, they have a midday bucket feed of carrots, swedes, apples and mangolds – put through the root cutter – or anything else in this line which is available. We sometimes give a mixture, but all these foods are introduced gradually to allow the digestive system time to adjust.

In winter, grass nuts are excellent for keeping up the milk yield. When buying these make sure that you can see bits of grass about half-an-inch long in the nuts. If the nuts have been made of powdered grass the ruminant cannot digest them and goats, being wiser than most, will simply not bother to eat them at all! Soak one nut in a little water and see the size it grows to and then you will not be tempted to give the goats more than they can comfortably accommodate, after they have swelled!

Browsings should be regarded as a titbit or condiment. It is a most uneconomical way of supplying green food because to do the job properly you would spend the whole day collecting it. The other disadvantage of relying on cut browsings fed in racks is that when left to their own devices in a wood the goats select what they eat themselves and they get a correct balance. You can quite easily carry something to them which, in the absence of a corrector, could seriously upset them. We once found our goats browsing on a rhododendron hedge. They had been out in the woods for several hours and were eating that hedge for about ten minutes before we reached them. Nothing happened except that one of them won a milking trial the next day! Another time a goat got out of the goat house and ate one leaf and we caught her and put her back in but that leaf nearly killed her.

It is far quicker and easier to feed a correct diet by growing your own green food, roots and vegetables. A good sample of oat straw is readily eaten by goats and makes a much cheaper midday feed in the late autumn and early winter. Its food value falls off after January.

The best and the cheapest concentrates are crushed oats. Flaked maize is expensive but helps with milk production and is good for putting on condition. Locust beans, decorticated groundnut cake and linseed cake are the best sources of protein which you can buy. Many people are reluctant to buy these because of the price, but we have found them much cheaper than dairy nuts. We used about 2cwt a month when we fed the latter but since we switched to the former we have used 1½cwt every *two* months. When you consider that dairy nuts contain maize, oats, beet pulp and sometimes even chicken manure, they are a very expensive way of feeding cheaper foods! Many people use bran in their ration but the only time we have used it, and that was years ago, was as an aperient. Continuous feeding of bran in the rations will cause a mineral imbalance which in turn leads to other troubles.

We feed the following mixture to our goats, which is balanced by all the other foods they eat. In the summer, and in the winter too if the hay is really tiptop, we do not feed the linseed cake in the morning ration.

Morning ¾lb crushed oats, 5oz flaked maize, handful of grass nuts, handful of locust beans, 1oz *Nutrichip* goat minerals, which is specifically blended for high yielding goats.

Winter Midday Root feed and a handful of grass nuts and crushed oats, (up to 3lb roots if they clear them up. In-kid goats are never given more than 1lb mangolds.)

Evening feed Crushed oats, flaked maize, grass nuts (same amounts as in the morning feed), handful linseed cake, handful locust beans.

We keep a mineral lick in each stall in addition to the minerals fed each morning. As we have an iodine problem here we add kelp but the amount varies according to individual requirements. Iodine has to be fed with care since too much will make the goat go off its feed altogether. One goat which has a tendency to periodical bouts of 'blowing' has been fed a teaspoon of brewers' yeast in her morning feed for months and we have since had no trouble with her.

One winter the goats had more than a ton of apples from various sources. We mixed the apples with mangolds in the root cutter, and these were fed in the buckets with the grass nuts.

Another good source of live food is greengrocers' waste. If you can find a supply ask if you can collect it weekly. It is best fed in troughs outside where the goats can help themselves to whatever they fancy.

May I remind you again that it is no use blindly following this method of feeding unless all other foods are available, otherwise you will run into trouble . . . even if that is only a reduced milk yield. If there is nothing else available you will probably need more concentrates but to my way of thinking that is very bad husbandry. A really good stockman can learn an awful lot about his own methods by just leaning over the gate watching his animals. By weighing and noting down all the milk yields you not only find out how much your goat is giving but you also find out if your management and feeding programme is successful.

5

GENERAL ROUTINE

'He must love doing that' (*Bob Martin*)

Routine work with your goats is roughly divided into daily, weekly, monthly and seasonal jobs. The daily routine consists of feeding, watering, milking, putting out to pasture, mucking out and dairy work. Weekly jobs cover changing your paddocks, electric fencing alterations, and attention to crops, including grass management and similar jobs. Monthly jobs include worm drenching, hoof paring, complete cleaning out of deep-litter beds, and building muck heaps. Seasonal jobs include haymaking, crop cultivations, muck spreading, crop harvesting and cultivations, and repairs to the goathouse.

Milking

Goats' milk is a very appetising and nutritious food when produced properly. Easily digestible, it is suitable for feeding to infants and invalids without pasteurisation and for this reason every care should be taken to ensure that it is produced under clean conditions in order to prevent contamination by harmful bacteria.

When you are starting up your goat enterprise you should arrange to do your milking outside the goathouse itself. Ideally it should be done in a room or stall separated from the goathouse by a door, with access to your dairy, which ought to be located beyond this stall all under cover, of course. Many people milk their goats on benches, but I prefer to sit on a milking stool with shorter legs than you would have for milking a cow. This way the goat stands on the concrete floor which is easier to keep clean than a wooden bench.

Most people find, when learning to milk, that the goat is distracted from their ham-fisted efforts if it is eating concentrates, and so the habit is fixed that the goat is fed while being milked. It is advisable to have either a yoke type of restriction or a collar to tie the goat, until you are used to the goat and can also milk expertly. After that it is seldom necessary to restrain a milker.

Strongly flavoured foods such as turnips, kale, swedes or silage must never be fed within four hours of milking, otherwise the milk will be tainted. It is best to give these feeds just after milking.

The dairy need not be large. All it needs is a water supply, a sink, a non-rusting rack to drain your milking equipment and a table. Obviously if you are going to sell your milk you will need a cupboard to store cartons, bottles, etc, a refrigerator and some type of bulk cooler. For small quantities of milk – two gallons and under – it is sufficient to stand or float the churn with the milk in a bucket of water under a very slow-running tap.

The most important part of producing goats' milk which is free from unpleasant taints is cooling, and this can not be done soon enough. Due to the fact that capric and caproic acids in goats' milk are in a different ratio from cows' milk, goats' milk will grow a very unpleasant flavour if kept at temperatures over 45°F for any length of time. I cannot stress this point too much. If you have several goats to milk and your dairy is away from the goathouse, it is best to put each goat's milk into a churn which is floating in running water, so that when you have finished milking all the milk is at tap temperature. You then take it to the dairy and strain it, put it in bottles or other containers and store it in the fridge or the deep freeze. If you strain the milk before putting it in the churn away from your dairy you must strain it a second time when in the dairy. It is impossible to have a strainer which fits on a churn so that dust cannot penetrate between the lip of the churn and the strainer. Many goatkeepers who thought this was the cleanest way to do the job have been puzzled by a sediment at the bottom of their jugs and bottles, but have found this vanished as soon as they strained the milk in the dairy.

Even if you only produce milk for the house you should observe these rules and never mix your dairy utensils with your kitchen pans and crockery. It may seem difficult when you are planning your enterprise, but it saves a lot of time and disappointments. It is so easy to get acclimatised to off-flavours in goats' milk, but it is a bad advertisement for goats' products when your friends and other visitors are subjected to this fare. The worst part is that the producer is probably quite unaware of its nasty taste and thinks his friends are either prejudiced or fussy!

Before milking, the goat should be tied up and fed and her udder and hindquarters wiped down with a damp cloth which has been soaked in a solution of a suitable udder wash such as *Iodip*, available from your local agricultural suppliers or a veterinary chemist. It is seldom necessary to do more than this as the well managed goat is a very clean animal. A small amount of udder cream such as *Cetavlon* or *Cetrimide* should be put on the teats and hands. The first four jets of milk from each side should be milked into a strip cup. This is a special container (holding about one pint) with a handle and a shelf of black plastic halfway down the inside; with it you can detect flaky or stringy milk which will indicate mastitis or some other fault.

The milk from such an animal should be kept separate from the rest pending investigations.

Most beginners will be wondering exactly how to milk, and probably hoping that their goat will be like the music hall cow that jumped up and down while the owner hung on! Once you have the knack, milking is easy, but you should be very firm with yourself and not use the 'bell ringing' style, which is the easiest way to get the milk but not at all good for the goat.

13 Ideal grip for milking

Hold the teat as shown (Fig 13). Never let the tips of your fingers touch the teat. Pressure is applied by the first finger in order to hold the milk in the teat, and then by the second, third and fourth fingers in turn in order to force the milk out. The grip is then relaxed to allow more milk to flow into the teat, and the procedure is repeated. A really good milker keeps her arms quite still, with little bouncing up and down. It is a good plan to practise milking with the finger of an old rubber glove with a pin-hole in the tip; you will soon find that the action becomes automatic.

When the milk stops flowing into the teats, gently massage the udder until there is more milk in the teats, and then milk out as before. Continue until the udder is empty. If you don't start stripping your goat with that finger and thumb action you won't miss it, and your goat will be better without it.

Milking should be done as quickly as possible, within

reason. If you milk too slowly the animal will 'hold back' her milk and this in time will cause the yield to fall.

When the milking is finished, weigh and record the yield, put the milk in a covered churn floating in running water and start on the next goat.

The object of weighing and noting down the amount is not just to tot up the yield of the goat. If you can arrange to have several weeks visible on one sheet you will be able to use the yield as a guide. Why did it go up or down? Is that goat getting wormy? Does she need a change of pasture? Would it have been better if she had stayed in yesterday because of flies, or rain, or high winds? In this way you learn more about how to produce milk and care for your goats than anyone can ever tell you.

Some people insist on milking exactly at twelve-hourly intervals, and others say that regularity is more important than the actual interval. Experiments with cows have shown that milking at twelve-hour intervals results in a higher milk yield after calving and a lower yield in the second half of the lactation, whereas milking with one ten-hour interval and one of fourteen hours results in a lower yield after calving but a higher yield in the second half of the lactation. The total yield in a single lactation will be the same in either case. These experiments also confirmed the importance of sticking to regular milking times. If winter milk is required, milking at ten- and fourteen-hour intervals is obviously the best system to use.

When milking is completed the churn should be taken to the dairy and the milk should be strained, bottled and placed in the refrigerator. (See Chapter 6.) It is a good idea to offer a drink of warm water with or without molasses at milking and feeding times. Goats seem to do better if they are left indoors with a small feed of hay for about an hour after milking, but in this herd the routine is varied according to the weather and other circumstances. Normally it is better not to put them out on a damp pasture. During hot dry spells the earlier the goats get out the more they can eat before being driven in to shelter from the flies. In hot weather we often put the goats out after the evening milking until it is dark.

The goathouse should then be dealt with straight away. If the goats are on deep litter any messy bits are removed, a fresh layer of straw put down if necessary, the gangway washed and water in the buckets changed. The house is then ready for the goats, and is also in a fit state to be seen by unexpected visitors.

Once a month throughout the year the goat needs its feet trimming. Possibly in very dry weather they grow more slowly, but nevertheless they should be looked over. Instructions for trimming are given on page 89.

Worming

Worming is a much neglected side of goat husbandry and worms are by far the greatest cause of loss of milk and condition. A great deal of nonsense is talked about worms by people who either ignore the problem, or who think that by keeping goats indoors all their lives, they are avoiding it. and I believe that goatkeepers who move away from what they call 'goat-sick pastures', do not know the meaning of pasture management.

Kids should be wormed at six to eight weeks if they are running with older stock. They should preferably be on clean pasture, one which has not had sheep or goats running on it for about twelve months and which has been grazed by other stock during the autumn, had the weed cut and then harrowed and rested during the winter. Otherwise running the mower over the pasture in the autumn, giving it a light spreading of well rotted dung, harrowing after rain and then leaving till the grass is about five inches high makes an excellent site for kids and lambs in the spring. Kids should continue to be wormed monthly until November and then twice more before the spring. After that the once-monthly routine should be continued. With milkers the same routine is observed, and the last worming before kidding should be not more than four weeks before she is due; if you are in doubt consult your vet. You can start again ten days after kidding. This is the most vital dose, since when a goat's resistance is lower, either after an illness or after kidding, the worm population increases rapidly, and a routine dose will keep this in check.

So much research continues into vermifuges and newer and safer ones are continually being put on the market that I think it would be silly to include names here. It is far better to discuss this problem with your veterinary surgeon. You need to change the vermifuge several times during the year in order to prevent the worms from building up a resistance to one family of drugs.

If the drug has to be administered by mouth (this is called drenching), measure the correct dose, put it in a small mineral water bottle with a sloping neck, and then gently pour it little by little at the back and side of the goat's mouth (see Fig 14). If the head is held too high and the liquid poured over the tongue

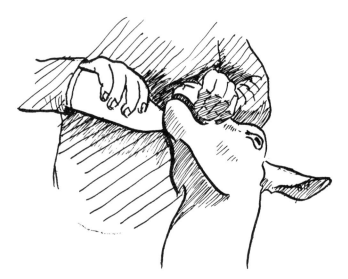

14 Correct position to drench a goat

at the back of the mouth it is liable to get into the windpipe and choke the animal.

If you always use the same vermifuge year in, year out, the worms become resistant to it and your worm control breaks down. Two or three times in the year you should give a different medicine. We are successfully using *Nemicide* at the moment and we find it very useful. It is administered as a subcutaneous injection (one given under the skin). There are several other vermifuges on the market but unfortunately these are made up in 500cc packs and are quite uneconomical for the small-scale goatkeeper to buy and keep, as a bottle would become unusable before even a quarter of it was used. The answer is to get your local vet to inject the goats or, if you can do the injection yourself, to get the single dose from your vet.

Annual inoculations against the *Clostridii* group of organisms are also advisable. We use *Covexine* which covers the whole group, and we do the injecting ourselves, but I would advise the beginner to start off through the vet. One of the worst of these infections is enterotoxaemia. This is a killer and there is no cure for it, so it is wiser to prevent. I believe that some farmers do not use this vaccine until they have an outbreak in their flocks. But the loss of one sheep in a flock of fifty is only one fiftieth of the herd whereas if you have only two goats it is half your herd.

6

DAIRY WORK

Dairy utensils, buckets and strainers must be seamless and preferably of stainless steel. Other metals tend to corrode very easily and aluminium gets pitted when left soaking in dairy detergent or hypochlorite. Enamel is completely unsuitable as it chips easily. Unfortunately stainless steel is very expensive but if you can afford the initial outlay it is more economical in the long run as it lasts a lifetime. Plastic is the least suitable since the surface is covered with microscopic pits which make it very difficult to sterilise, and you quickly get rather smelly containers which can spoil the flavour and the keeping quality of your milk. Filter pads should always be disposable.

Regulations concerning the sale of goats' milk and its products are few. If you sell direct to the consumer there are none. If you sell to a retailer you have to have your name and address on each container. You must also have the quantity marked on the container unless you price it per carton or bottle, and you must state that it is goats' milk, or a goats' milk product.

Plate 15 Contented goats finding their share of wild herbiage
Plate 16 Author (left) and partner with Cothlan Dewdrop and Cothlan Sally

Plate 17 British Alpine class being judged at Alresford Show, Hants
Plate 18 'High level milking parlour!', Blandford Fair, USA

This all sounds very simple but there is one snag. If someone takes a sample of your wares to the public analyst and it is found to contain harmful bacteria, too much water or some other abnormality, your premises will be inspected. If your dairying methods fall short of those laid down for cows you will have to stop selling your produce until you have brought your dairying up to these standards. This will mean that the walls, ceilings and floors of the milking room and dairy will have to be faced with a suitable washable surface, among other things. It is for this reason that I have suggested that you have a separate milking room and dairy from the outset. It is easier to plan it right from the start than to add it later.

Milk containers are becoming a very great problem. If you have only a very few retail customers the easiest milk containers are those squarish upright glass jars which contain fresh fruit juice. These usually hold about 33 fl oz. The wide caps are made of metal coated with a surface which can be sterilized. They also pack very well into the fridge. In spite of the Ministry of Health's ruling that all returnable bottles must be washed and sterilised in a bottle washing machine, when questioned, a Ministry official told me that provided the sterilising was done efficiently and with the correct hypochlorite and dairy detergent and provided the bottles were used for retail customers *only*, he could see no objections. But you must *never* use narrow-necked squash bottles. Normal milk bottles with a metal foil cap are becoming a problem. The bottles are easy enough to obtain, and so is the capper, but the caps can only be purchased in lots of 10,000. The last time we ran out we went to an agricultural merchant who sells them and asked him if he had sold any to a customer who had complained at the huge quantity. He had. So we contacted the farmer and bought half his caps (by weight!).

Pint cartons suitable for the sale of fresh or deep frozen milk are now very expensive but fortunately Lakeland Plastics Ltd (address on page 160) have now produced heavy white plastic bags overprinted with a blue goat design and all relevant information, and a matt panel for overstamping with the producer's name, in pint, quart and half gallon sizes. We have used the pint bags for some months now and find them very satisfactory; they take up less room in the freezer and the customers like them. There is a full range of cartons, etc, to match. Lakeland Plastics will deliver your orders to the nearest show to save postage. They supply electric heat sealers and a milk-bag holder for use when filling. The special waterproof FP ink for use on plastic may be purchased from Express Rubber Stamp Services Ltd (address page 160).

Handling milk

After milking, cooling and straining the milk is ready to be bottled. For liquid sales this should be done at once and the containers sealed and then placed in the fridge. This also applies to milk for home consumption. For deep freezing, the sooner the milk is in the freezer the better and in any case it should be in the freezer not more than twenty minutes after milking since the object is to get it frozen before the cream starts to rise. Freezing is more successful with individual pints, though milk from many semi-commercial herds freezes in gallon containers, which are later delivered to dairies which specialise in goats' milk and yoghourt.

The storage temperature for deep frozen milk is −10°F. If at any time the temperature is raised to 15°F or a little higher (eg while delivering to a wholesaler) and it is then reduced to −10°F again, the milk when thawed out will be flaky and unpleasant to look at. This does not appear to affect its feeding value but it is obviously undesirable. Deep frozen milk should be allowed to thaw out naturally at room temperature. It can be put in a fridge to thaw out; you will find that the milk will last about a week this way.

It should be remembered that milk is plentiful in summer and only those producers who supply winter milk will get the manufacturers' preference in the summer. Any goatkeeper who wants to sell his milk should base his summer milk sales on the amount he produces in winter and should be prepared to have other methods of dealing with the surplus milk in summer.

WASHING AND STERILISING

All utensils should be rinsed in cold water and brushed with a proper dairy brush. They should then be put in a solution of dairy detergent and dairy hypochlorite. *Deosan* make a suitable one and instructions are printed on the containers. These are available at agricultural and farm suppliers or agricultural and veterinary chemists in gallon containers. It is useless to use domestic detergents. While hot water is more suitable most of these solutions work well in cold water. The utensils should be brushed in the solution, rinsed in cold tap-water and placed upside down on the rack in the dairy to dry. Never wipe dry with a cloth. Milk bottles can also be sterilised in this solution. The solution can then be used to wash the table and floor of the dairy and milking room. If you have to process your milk in the kitchen you should never sell it.

Goats' Milk and its products

Goats in the British Isles are free from brucellosis (*abortus*) and tuberculosis and that is why the ministry does not require the milk to be pasteurised, nor the herds to be regularly tested. Malta fever (*melitensis*) has not been known in this country since 1922 and that case was recorded in a goat which came from Malta.

The fat globules in goats' milk are much smaller than those in cows' milk and this makes it easier to digest. All milk curdles when in contact with the gastric juices in the stomach; cows' milk curdles into a solid lump, but goats' milk curdles into flakes and this makes it far easier to digest. Many conditions seem to improve when goats' milk is included in the diet, which seems to help the digestion of all other foods as well.

Frozen Milk   The milk is frozen in pint cartons and must be allowed to thaw out naturally, taking about twenty-four hours in a cool larder. Provided no attempt is made to refreeze this milk it will store satisfactorily for ten days in an ordinary refrigerator.

Clotted Cream   This is similar to cows' cream but has no colour unless added by the dairyman. It can be used and stored in the same way as cows' cream. Freezing is not practicable.

Butter   This also is colourless but otherwise similar to cows' butter. Being a fresh food with no preservatives this product should be eaten quickly and should not be moved in and out of the fridge as this hastens souring. It can be frozen.

Cheese of the cottage variety is also colourless and should be treated in the same way as other cottage cheeses. It freezes very well.

Yoghourt   This product is made entirely of goats' milk with the addition of live cultures and contains nothing else, not even dried milk. It stores satisfactorily for two or three days in a fridge, but must not be frozen.

Recipes

EASILY MADE YOGHOURT

This can only be made from pasteurised milk which is free from antibiotics.

Utensils   1 Dairy thermometer, 1 large basin which fits neatly into the top of a saucepan, or if making smaller quantities, a quart vacuum flask.

Method   To pasteurise new milk heat to 180°F–190°F and keep it at that temperature for $\frac{1}{2}$ hour. Cool to 118°F. Add 1oz of a good *dairy* yoghourt to each pint of milk. Stir well.

If you are using the basin method, put this cultured milk in the basin and stand it in a saucepan filled with water at 118°F. Place at the side of the cooker and cover with lid. It should take about 5–6 hours to become yoghourt. If you want to test it for acidity, take some from just below the surface and not from the top.

To make in quart vacuum flask: heat 1¾ pt milk and treat as directed above. Put in flask at 118°F and add culture to fill the jar. Stir well. Cork with sterilised plastic cover and leave for 6–24 hours according to the degree of acidity liked. Cool and stand in the fridge uncovered overnight. Next morning stir to mix well and put in pots.

To Make Fruit Yoghourt, use pale moist brown sugar for sweetening.

Raspberry flavour, use fresh or frozen fruit. 1 heaped tbsp raspberries to each 1pt yoghourt, 2 heaped tbsp sugar to each pint. Blend sugar and fruit well and stir in yoghourt.

Strawberry yoghourt needs 1 small cupful fruit to 1pt, with 2 tbsp sugar.

Apricot Soak dried apricots, cook them and put through the liquidiser. I use 2 tbsp of this purée and 2 tbsp sugar to 1pt yoghourt.

You can make a pleasant yoghourt with a blackcurrant cordial. We like one sweetened with honey and flavoured with fresh lemon juice.

Or add yoghourt to salads, mixed with chopped chives and diced cucumbers, or cucumbers and a dash of mint.

LEGAL ASPECTS OF YOGHOURT MADE FOR SALE

In the UK the Food and Drugs Act 1955, the Trade Descriptions Act 1968, and the Labelling of Food Order 1953, (Regulations 1967) apply. Cartons must be overprinted with the name and address of the producer, and weight or volume of contents, and constituents listed. There is no legal definition of yoghourt. If made from whole milk it should contain at least 3 per cent butterfat and 8.5 per cent solid not fat, but if labelled low fat then 1–2 per cent butter fat might be expected as opposed to 0.05 per cent butterfat in a so called fat-free product. Fruit yoghourt would be expected to contain 15 per cent fruit pulp and sugar to the extent of, at most, 5 per cent.

YOGHOURT CULTURES AND CHEESE STARTFRS

If you want to make yoghourt and cheese for sale, you need to make and keep going your own supply of cultures. We find the freeze-dried cultures are far the best and most convenient as you can store them indefinitely until needed. These can be purchased from Hansen's Laboratory Ltd (address on page 160).

For yoghourt you need the ordinary yoghourt culture. There are instructions for making your cultures sent with the phials, but all liquid measures are in litres which we found difficult to follow and we tried different ways before we hit on the simplest. We use a quart vacuum flask and pasteurise 1qt milk as for yoghourt and cool to 118°F. Put it in the flask, sprinkle in the whole of the contents of the phial and stir well. Seal with the cork covered with a sterilised plastic bag and stand in a warm cupboard for 24 hours. It is now ready for use. One 'brew up' of culture like this lasts us for 2–3 months as you merely use fresh yoghourt for the new batch. This all depends on how careful you are with sterilising the utensils.

TO MAKE CREAM CHEESE YOU NEED THE 'ORDINARY LACTIC FERMENT, FREEZE DRIED' CULTURE.

To Activate the Cheese Culture follow the directions supplied, only using 1qt of milk in the quart flask. Once you have made this culture you have to re-make it from your existing culture every other day. To do this heat 1¾pt milk as directed for cheese (see below), put in flask and stir in ¼pt of old culture. Seal and stand for 24 hours. When I have had too much culture hanging around I have dumped the lot in the cheese brew and it makes a very nice lactic cheese. You can also use a little of this as a starter when making butter.

For a group of goatkeepers living in the same area and all wishing to make yoghourt and/or cheese, it would be economical to share the first culture. Divide it out after the propagation and then go your own ways.

When purchasing a dairy thermometer, don't buy one in a wooden holder as these are expensive and *quite* unsuitable.

The glass ones are inexpensive and can be purchased from most agricultural suppliers.

Utensils Two 2 gall plastic buckets, 3 double muslin bags, or Harrington squares made into bags for straining, dairy thermometer, ordinary rennet, cheese starter (as discussed above), salt.

Method Heat 1½ gall milk to 85°F, stir in 4oz cheese starter which you have prepared the previous day, cover with cloth and stand in a temperature of 80°F for 24 hours – the airing cupboard is a good place for this. The mixture should now smell cheesy and be beginning to thicken. Stir 1tsp rennet in ½pt cold water and add this to ½gall of fresh, tepid milk. Stir well and add to the 1½gall of the cheesy brew. The whole should now be quite cool. Cover again and stand in an even room temperature (56°F) for 24 hours.

It is most important to follow the next part of the directions exactly as otherwise half the cheese will flow through the straining bag. Arrange your cheese bag in the bottom of a 2gall bucket and peg the sides round the lip of the bucket. Take a small mug and gently bail the cheese mixture into the bag. Do not tip, just let it flow into the bucket. When all is transferred, gather the top of the bag and then firmly and gently lift up so that it swings free to drip. Leave for 24 hours.

Change your cheese bag. This is a messy business but you soon find the easiest way of doing it. Let it drip for 24 hours.

Change bag again and let it drip for a further 24 hours. The cheese should now be dry enough. If not, hang it for a further 24 hours in a clean bag. Turn into a basin and stir in salt to taste. Pot in plastic tubs.

We find this cheese is most successful frozen and we are still eating the August cheeses in February. You can add all kinds of flavourings to your cheese – herbs, chives, pineapple, minced ham, etc.

CLOTTED CREAM AND BUTTER

Goats' milk cannot be successfully skimmed by hand, since it takes so long for the cream to rise that the resulting produce usually gets a 'goaty' flavour. For successful skimming a separator is a must. Heat the milk to 103°F and put through the separator. Collect the cream in a shallow bowl. Place the bowl over a pan of water and simmer until a crust has formed right over the cream. Cool and put in fridge.

The cream is now ready for use. After four days it should then be ripe enough to make butter. If you have any cheese starter put a teaspoonful into your cream. Add a dash of Annatto butter colouring. Now scrub your hands well and hold your working hand under the cold tap until it is very cold. (If preferred and you have an electric mixer, use the K beater.) Stir until the colour begins to change and the texture starts to granulate. Add half a cup of 'breaking' water and continue to stir. If very stiff, add more water. When the cream separates into liquid and small granules, strain it through an ordinary nylon strainer. When the strainer is half full run the cold tap through it, until all the buttermilk has disappeared. Shake out water and tip the butter into a clean bowl. Repeat till all butter is washed. Add salt to taste and mould up into a lump, squeezing out water, and 'pat' into shape required. The butter will freeze well.

Separators are once again on the market but are very expensive to buy so it might be worth experimenting with setting cream to rise in the fridge, so that you would be making clotted cream in the real old-fashioned way. Strain the milk into a shallow pan which will fit into the fridge and on top of a saucepan. Cover the pan with transparent 'cling wrap' and put in the fridge for about twenty-four hours. Place on the top of the saucepan filled with water and bring this to simmering heat. Keep at this temperature until the cream has a crusty look. Cool down as quickly as possible, cover, and return to fridge. This should be ready to skim after twenty-four hours or a little longer. A proper skimmer is a perforated disc with a small handle. Skim off the cream into a smaller bowl, leaving the skim milk for puddings, stock feed, etc.

Dairy Work

I understand that the difference between the old Cornish clotted cream and Devon clotted cream was that the former was heated over charcoal in shallow pans as soon as it was drawn from the cow, while the latter stood for twenty-four hours before being heated over simmering water.

7

THE MALE GOAT

Keeping a male goat is an undertaking which must not be taken on lightly. It is no job for the timid, the weak or the novice. A fully grown male can weigh upwards of 250lb. He is apt to be a house and fittings basher, and some have a great aptitude for literally 'raising the roof'. You are lucky if you have a male who respects his fences, and many do not care too hoots for an electric fence.

From August until at least February the males spray themselves and exude the most awful smell which they rub off onto anything they can get their heads near – other goats, their house and, best of all, *you*. It is almost impossible to remove this smell, but like all other 'off' flavours, you do get used to it, and it is this which gives goats and goatkeepers such a bad name, or should I say smell! If the male is kept with the milkers everything takes on his aroma, including the owner, the goats, the goathouse (your *own* house) and especially the milk. This is one reason why the goat's reputation for being a smelly animal and producer of nasty-tasting milk has grown up. This is still a universal belief and it is due to goatkeepers' carelessness that this idea is as prevalent today as it was sixty years ago.

There are ways of avoiding the smell. When dealing with the male wear special clothing reserved for the purpose – an old plastic coat and rubber gloves. These are best kept in an outside shed! If, by mischance, you do get the smell on your hands, clothes or in your vehicle, there is a simple way of removing it. Beg, buy or borrow a bee smoker from a beekeeper, discover how to use it, then hang your smelly garment inside your smelly car and shut all the windows and doors. Through a crack which can be shut, pump all the smoke you can get out of the smoker into the car, block up the crack and wait until all the smoke has vanished. The car and its contents will smell like burnt cardboard, but when you wash them the male smell will have gone. Similarly, if the smell is on your hands rub them in the smoke from the fire or bonfire and it will go.

All male kids unless specifically ordered or wanted for breeding, should either be castrated and reared for meat or destroyed at birth. Now I know very well how sweet they are when they are little and what lovely pets they would make! But for how long? Six weeks at the most. And who would want a pet which needed four pints of milk in four feeds a day? There is more cruelty caused to kids by sloppy, sentimental people who won't make the effort to destroy such pretty little creatures, so they give them away 'to good homes'. And what happens. . .? Well, for a start the goats probably grow horns which can, even unintentionally, cause nasty accidents. Then the males grow big and rough, no longer pretty little pets, and are given away to some other unsuspecting soft-hearted person. And so it goes on, each person in the chain forgetting the episode as soon as the animal leaves, but the poor goat, mostly thoroughly underfed and possessing unpleasant habits which nobody understands, goes from bad to worse and *nobody* ever seems to have the guts to have him put out of his misery and humanely destroyed. Even the police and the RSPCA, in my experience, fall flat on their faces when confronted with a problem like this.

If you do propose to keep a male, you need a very strong house at least seven feet high inside, with a good well-fenced run, or series of runs, in which he has plenty of room to dash about, and from which he can watch you at work, and the goats at pasture. We like our milkers or kids to be able to graze in an adjoining field when possible; their presence keeps the male contented.

It is a good idea to be able to get into his house without coming into contact with him, so that you can feed and water him without having to get dressed for the occasion. This is all a matter of planning, and it is well worth thinking it out before starting, as it makes life much easier later.

There is no point in keeping a male for a small herd. It is far less expensive to take your goat to a good male, however much the stud fees cost, than it is to care for a male goat and feed him properly for 365 days of the year. There are advantages, of course, but these are far outweighed by the disadvantages, and, until the novice goatkeeper has had quite a lot of experience with milkers and rearing young stock, he would be well advised to steer clear of keeping a male goat. Nowadays there are plenty of stud males available within reasonable distances in most parts of the country, with good production records behind them. Stud goat lists are available from the British Goat Society and from most affiliated societies too.

Feeding the male is similar to that of the female, but being a larger animal he needs more food. As a kid he should get four pints of milk daily at first, reducing to two by the autumn. He needs plenty of hay, green food (and especially if he cannot go out to graze himself), roots and kale. He should *not* be fed mangolds or sugarbeet. It is best to give him rain water to drink.

The male needs his feet trimmed regularly and this can be an energetic and unpleasant occupation. He also needs worming and inoculating at the same intervals as the female.

The Male Goat

The smell of the male is exuded during the 'rutting' season (August to February) from scent or musk glands normally situated on the crown of the head between the horns or the polls which grow on naturally hornless goats. The unpleasant habit they have of spraying their head, shoulders and front legs during this period helps to spread the odour, ably assisted by his love of rubbing his bedewed head on everything and everybody. Of course, this is most attractive to the female goat, and can be smelt by her from miles away. I was once down in the village, over a mile away – down wind that day – when somebody remarked that there was a curious smell about which they couldn't identify. It had been around all day, so it wasn't me, but it *was* old 'Midas'!

Some things can be done to lessen this disadvantage. Keep your male on clean dry straw bedding. You can also have the musk glands on male kids cauterised which, if it doesn't cure the smell, at least lessens it. When you do his monthly hoof pare wash his front legs, neck and head at the same time. Often too much enthusiastic spraying causes the hair on these parts to fall off or the skin to form scabs. Use a good cattle shampoo and oil the sore parts afterwards. It does help to improve his appearance when newcomers to goatkeeping come visiting with their goats for his services.

A male kid should be given at least one bottle all through the winter and he should be encouraged to keep eating his concentrates. Some males go off their feed during the rutting season and this is a bad for them. Plenty of kale, grass, lucerne, oats, vetches and *trifoleum*, etc, all the year round will get him into good eating habits and, if he does go off his concentrates, you can keep him going on good greenstuff and hay. Whatever you do, keep him contented and he will probably go on eating throughout the winter.

There are different schools of thought on the subject of how many goats a kid should serve in his first season. The most contented males we have reared have been the ones we have used as often as necessary, but we do limit a kid to one service a day. If male kids do seriously go off their food then you will have to restrict the number of goats they serve.

Generally speaking it is advisable to keep good tempered males, because they do pass on these qualities to their off-spring, and there is nothing easier to deal with than a contented herd of milkers.

Male kids can be fertile from 3 to 4 months so should never be reared with the herd or with female kids. We knew one female which kidded to a companion when they were only 10 months old.

Ministry of Agriculture Legal Requirements for the Castration of Goats Up to one week old castration is permitted by the use of a rubber ring or other device to constrict the flow of blood to the scrotum by an unqualified person without anaesthetic. Up to two months castration by an unqualified person without the use of anaesthetic. Two to six months as above, but with the use of anaesthetic. Six months and over should only be carried out by a veterinary surgeon with the use of anaesthetic.

Deodorising male kids

Most of the smell produced by the adult male is produced by cell masses located towards the middle and behind the horn buds in the young kid. When disbudding the kid the whole of this area should be shaved before the operation. It will then be possible to see these glands, which are yellow masses about the size of small match heads. These may be burnt off at the same time as the horn buds are treated.

Female goats do not seem to be attracted to deodorised males!

8

BREEDING

In the wild goats normally come into season in the autumn and produce their young in the spring. Nature usually organises this to coincide with the annual flush of grass. The dam produces enough milk to feed her kids until they are old enough to fend for themselves. Those females who produce too little milk rear puny kids which do not survive, so that the family soon dies out. Other abnormalities are eliminated in the same way. Domestic goats, over the years, have been bred to produce much more milk than wild goats. During this process breeders have been able to rear many more kids than would have survived in the wild.

Goats should be mated for the first time when they are yearlings. Novice goatkeepers should stick to this rule until they have had several years of rearing and feeding stock and understand the subject. Kids which have grown well can be mated in January or February to kid during their goatling year and experienced breeders sometimes do this, but it is not usual.

Milkers, ideally, should be mated every other year. Thus a householder with two goats will mate one and keep the other in milk for twenty-two months, mating her the following September. This will keep a continuous supply of milk for the household. The same applies to commercial herds. Half the goats one year, and half the next, keeping enough milk throughout the year for the customers.

Goats will only allow the male to serve them when they are actually in season. The period of oestrus lasts for one to three days, and is repeated again after twenty-one days. This cycle continues from about late September until February, unless the goat if successfully mated 'holds' to the service. Oestrus is characterised by a general uneasiness of the goat. She becomes much more 'talkative', often bleating continuously, and she wags her tail in a very distinctive way. There is often a swelling of the vulva and a mucous discharge. If let out with other goats you will find that she will mount their backs and allow them to mount her. Sometimes the milk yield falls and sometimes it goes up; you should always watch out for this on your milk recording sheet.

You should make arrangements with the owner of the stud goat in advance and he will tell you when you should notify him that your goat is ready for service. The normal practice is to ring the stud owner at some specific time on the morning that your goat has come in season and arrange when to go to the male. Over the years we have found it best to unload the female and let her get her bearings for a minute or two. Our male is kept some way away from the gate, so that the goat can find out in her own time that there is a male about. If she is ready for service she will walk willingly towards his pen and will wag her tail on approach. When let in, she will allow him to sniff her all over and after a very few minutes he will mount her back and mate her. You can usually tell when it is success-ful because the male throws his head back and the female humps her back after service. We have found that it is as well to let them stay together until he serves her a second time, which is usually between five and ten minutes later. After that they do not mind parting and we find, unless there is something wrong, that the goat holds to the service.

Stud goat owners issue you with a service certificate at this time and you also pay the stud fees. If the goat comes in season three weeks later it is normal practice for the stud owner to let you have a free mating. If after this second mating she returns you should seek the advice of your veterinary surgeon.

The normal gestation period is from 145 to 155 days and a very simple method of arriving at the date on which your goat should kid is to count five months forward and subtract one day. If she is mated on 19 September, she should kid around 18 February.

Care of the in kid goat

When you have got your goat in kid – ie she has passed the third or sixth week after mating without coming into season again, and her milk yield is beginning to shrink, don't give her less to eat. She needs the good food you gave her for milk production to feed her growing kids.

People have different ways of telling whether their goats *are* in kid. We have found that somewhere around two to three months into the pregnancy the skin around the vulva becomes puffy and soft if the goat *is* in kid but that it does not change if the goat is swelling with a false pregnancy (cloudburst, see page 136).

Her feet need attention and should be cut the *last* time before kidding, about four weeks before she is due. Even if they grow too long after this cutting, they should not be touched again until at least two weeks after kidding.

HOW DO YOU CUT FEET?
Fig 15a shows what a properly trimmed foot should look like. The white part is the 'nail' which needs trimming, and the shaded part is the soft part which, in theory, should not be cut,

but which often grows a heel which needs to be trimmed off, as shown in Fig 15b.

The 'nail' of the hoof is the part that grows fast and needs regular cutting. It should be cut flush with the soft part of the foot (See Fig 15c). Sometimes this 'nail' grows so fast that it turns under and completely covers the soft part. Don't take a great sweep to cut this off in one go. It is much safer to cut off small bits until you can see what you are doing. In all cases of hoof trimming, run the animal on wet grass for about half an hour before you start, as this will soften the hoof. Foot-rot shears purchased at an agricultural supply merchant or a veterinary chemist are a great boon and very easy to use.

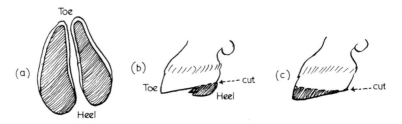

15 Trimming feet

LICE

These are very common winter visitors. All goats should be dusted with an insect powder about four times during the winter, and in-kid goats should be dusted two weeks *before* they are due. Louse powder usually upsets kids so that it should be given a chance to work off before they are born.

All goats must not be groomed during the winter months because the brushing will pull out their protective undercoat.

EXERCISE

It is most important to keep the in-kid ladies well exercised, whether they like it or not. It is no use turning them out into a yard or a muddy field, because their Union does not approve and they just stand and shiver. Take them for a walk (ten minutes is enough) round and round the yard if you have nothing better, but do not let the goat persuade *you* to take all the exercise! Make sure that the in-kid goats do not have to walk round awkward corners, and try to persuade them to stop squeezing through small holes! Keep them out of the way of any males in the herd.

Kidding

The yield of most milkers rises for about six weeks after mating and then gradually falls, until at about six or eight weeks before they are due they usually go dry. Some goats are impossible to dry off and care should be taken with them to see that they get extra concentrates at this time to compensate. It is also most important not to cut down the concentrates when your goat goes dry. She is feeding her kids; during the last six weeks of the pregnancy they put on about seventy per cent of their weight.

You should make a careful study of your goat's rump and hind legs six weeks before she is due, so that when kidding draws near you will quickly notice when her bones begin to unlock and the hollows appear on either side of the tail.

About two weeks before your goat is due to kid you should prepare a stall for her. It should be well scrubbed with disinfectant and, when dry, bedded down with short straw. This is better for the kids as they are not so likely to get tangled up. The drinking bucket should be too high for the goat to drop her kids in where they could drown. The gate should open outwards.

About two or three days before she is due she can be moved to the kidding stall, if it is in the goathouse. Provided she is well and not being worried by the other goats she can still go out to pasture with them. Take care that she does not have to squeeze through narrow gaps when she is heavily in kid.

As mentioned above, her backbone from the hips to the root

of the tail should start to ridge up, and just before kidding hollows should appear on either side (*see photograph, page 35*). Her udder should have started to grow and the rest should gradually fill until just before she is due to kid it will appear to be filled and look shiny. Sometimes a goat will reach this stage long before she is due to kid and this will mean you will have to 'pre-milk' her. It is always a difficult decision to make because there are so many old wives' tales about it, but if you do not pre-milk in such a case you run a very great risk of mastitis developing, and that is a long, difficult and sometimes impossible condition to cure. By pre-milking you do run the risk of milk fever, but this can be cured much more easily.

If during the last half of pregnancy your goat goes off her food and seems to have painful feet and is generally upset, test her urine for ketones. To do this you obtain a bottle of *Acetest* pills (for humans) from the chemist. Take a sample of the goat's urine and put a drop on the pill. If it turns any shade of mauve or purple you should contact the vet at once. At this stage the trouble is called pregnancy toxaemia and it needs to be dealt with at once otherwise permanent and incurable damage is done to the liver, which reduces the goat's milk production for the rest of her life. About eight weeks before our goats are due we test them as a matter of routine, at first weekly and later twice weekly even if nothing seems wrong with them. We continue this testing until about four weeks after kidding if they are normal, but if they have either pregnancy toxaemia before kidding or acedosis after kidding we continue until four weeks after the last positive test. In all cases the trouble must be dealt with by the vet.

Signs that your goat is near kidding are numerous. She may walk to the door of the goathouse when you let the others out, but then hang about or return to her stall. She gets more talkative and bleats whenever she sees you. You will find her digging and pawing about in her bedding. Sometimes she lays her head on her backbone as if it is uncomfortable. She appears to have gone down in size. Her respiration rate increases the nearer she gets to kidding. At this stage there should be a colourless discharge and as she gets nearer to kidding she makes a deep grunting bleat everytime she strains. And yet it is still difficult to tell when a goat is actually going to kid, even after more than thirty years of watching them. You can have all the foregoing signs, except the grunting bleat, and your goat is down and with bated breath you wait for the real heave. There is a pause in her breathing, she makes a grunting heave . . . and up comes the cud which she chews contentedly. And so on. In fact this year, in spite of everything, we missed all our kids being born. As it happened all were fine but a goatling with a multiple birth, however natural, needs help with the kids. If the caul is very thick the goat will not break it and unless you are there to do it for her the kid dies.

When she really gets down to kidding the colourless discharge changes to a thicker white and a shiny balloon-like bag appears at the vulva. This should break and you will see, in a normal birth, two front feet and, a little further back, the nose. Provided your hands and arms are well scrubbed, your fingernails cut short, and you have smeared on some obstetric cream, it does not hurt to give a little assistance at this stage. When the goat strains you can gently pull the feet and this helps to bring the head forward. If she stops straining, stop pulling. Once the head is through the rest of the kid slithers out easily.

If a goat is straining properly at fairly close intervals, and then stops altogether, a beginner should get help, either getting advice from the vet or actually getting him to see the goat. Some people have a natural aptitude for obstetrics and others have none. To all inexperienced goatkeepers I would suggest the vet, who will teach you how much you can do yourself and later will advise you if you get into difficulties. It is penny wise, pound foolish not to get expert advice as soon as anything goes wrong.

Figures 16–20 show most of the usual presentations and the captions tell you how to deal with them. It is wiser to get

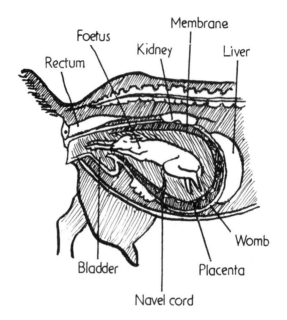

16 Normal birth presentation

17 Normal presentation for twins

18 Abnormal presentation: kid must be pushed back and hind legs brought forwards, then the kid is delivered backwards

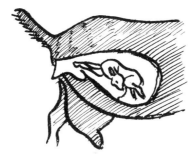

19 Abnormal presentation: push the kid back and gently bring the head into the normal position on its legs for delivery

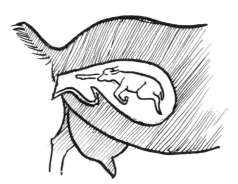

20 Abnormal presentation: kid must be pushed back and the legs straightened into normal birth position

someone else to hold the goat while investigations and manipulations are taking place. Otherwise the goat should be tied to a wall by a collar and lead. If you have to put your hand inside, be sure your nails are cut short and scrubbed clean and hands and arms well washed before anointing with obstetric cream. The rear of the goat must be well washed too. Keep your fingertips close together.

In a normal birth the second kid should follow the first after about ten minutes. The afterbirth should follow from thirty minutes to four hours after the kids have been born. If it is still there and does not appear to be moving at all, contact your vet.

If you are not sure if there is another kid to be born, put your hand firmly below the goat's belly and just in front of the udder and push up gently. If you feel a solid lump then there is another kid. If not, and the afterbirth is taking a long time to come away, push on this spot, lifting the belly as far as you can, and you will see the afterbirth coming out a bit. It sometimes speeds things up. You can get the same results by using a towel as a sling and lifting the belly that way. A drink of oatmeal gruel, made from a handful of coarse oatmeal and two pints of boiling water, sometimes speeds the arrival of the second kid. This gruel should be well stirred and cooled down to a pleasant warmth with cold water. Nearly all goats drink this greedily during parturition.

Once the kidding is over and the afterbirth has come away, the goat should be washed down with warm water and disinfectant, and then dried. The bedding should be cleared out and changed, a little ivy and kale and lots of hay put in the rack and a bran mash in her bucket, and she should be left to sleep it off. Make sure that the kids can suck before you leave them.

Be very careful not to give the goat too much to eat in the way of concentrates for about a week after kidding or she will go off her food and it will be difficult to get her eating again.

However you decide to rear the kids later, it is an excellent plan for the novice to leave the kids with their dam for four days. Leave them all in the kidding stall for those four days however fine it is. It is so easy for her to get a chill in the udder. By the end of the four days she will be longing to go out with the herd, and she will not mind parting from her kids so much. Leaving the kids with the dam for those few days means that her udder will be kept soft and this saves you the bother of doing it. Also they get a little milk as often as they wish which is much simpler than if you have to give it to them.

On the day when you let the goat go out with the herd, and when she is well out of earshot, remove the kids out of hearing of the goathouse. Clean out the kidding stall, and when you put the goats in let the dam go back to her own stall. If she cannot smell the kids in the kidding stall she won't fret half so much. If you can keep the kids right away from the herd for two weeks you will find that it is quite all right to put them back then. They will not try to suck from their dam and she will not let them.

After your goat has kidded it is most important to get her eating and drinking well. She should be given the very best green food you can find for her while she stays in with the kids. As goats nearly always kid in the early spring, it is not always easy to do this; but it is well worth the effort. She can be tempted to drink warm water, with or without salt, or you can add molasses to it at the rate of about a cupful to a gallon. Some goats will drink their own milk, if it is offered to them, for the first few days and this is not a bad idea, provided you stop it after that. Some goats love milk, but there seems little point extracting it from one end only to pour it back down the other – unless the goat is ill. It is true that they probably milk that little bit better if milk is fed back all the time but it is a costly way of getting an extra pint. It is also against the rules to do this at shows, and this puts goats who are fed milk at home at a great disadvantage if they compete in milking trials and their owners stick to the rules and stop feeding it!

After kidding you should feed the best hay you can find. If you have a variety, try to save the best for this time.

9

KID REARING

Selection

You should only keep kids from goats which you know to be good, long-lactation milkers, ie goats which will milk well through two winters without kidding, which is called 'running through'. Some show herds make a practice of kidding their goats every spring so that they have a crop of kids to sell and so that the goats will be in full milk when the show season and milking trials start. I think this is an awful mistake because it hides from records the knowledge that goats have or have not the ability to run through. In doing this there is a great risk that we will breed out this ability.

Kids which have the following abnormalities should be destroyed at birth: *Hermaphrodites* or kids with mixed sexes. This is not uncommon in goats particularly if the kid's parents are hornless. These kids appear to be female but when the vulva is examined it will be found to be larger than normal and the tip at the bottom will have a small pea-shaped body which can be felt on examination. If your goat has had a long and difficult kidding genital organs on both male and female kids appear to be larger than normal and also pinker. In this case leave the kid for two days before making a decision (Fig 21).

Overshot or undershot jaws In both cases the kid's teeth will not meet the pad at the top of its mouth, thus making it impossible to graze properly.

Supernumerary teats These are an inherited malformation. They appear in different forms and the breeder should form a definite policy, right from the start, about this problem. In this herd we always discard kids with any teat abnormality and for some years past we have been free from this trouble. But it is a subject of much debate amongst breeders because it appears unexpectedly on kids which may otherwise look perfect and be from exceptional breeding lines.

There are a number of such malformations (Fig 21). An extra-blind teat is one which has no gland above it and many breeders snip these off soon after birth. If the teat has a

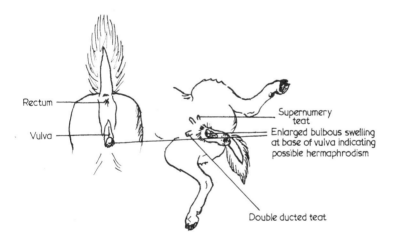

Rectum

Vulva

Supernumery teat

Enlarged bulbous swelling at base of vulva indicating possible hermaphrodism

Double ducted teat

21 Malformations

mammary gland above it, removing the teat will cut off the outlet for milk when the animal comes into production. This will cause trouble which could result in mastitis. The kid should be culled. Some teats have more than one orifice and others have two ducts and kids with these malformations must also be culled. Some normal teats have a branch teat sprouting out of it which makes milking a very messy business. This kid too should be culled.

Kids with unsound legs, crooked backs or other structural faults should be culled at birth. Of the kids which remain after this examination save only the best for replacements. It is a good thing at this stage to make a resolution that what is not good enough for you is not good enough for anyone else either – unless it is a mismarked kid. Selling poor animals hurts you as much as the buyer and it is a very bad thing for goatkeeping in general.

There is little value in keeping records beyond the third generation. If you try to select your animals for too many things – production, pedigree, conformation, udder, colour and markings, etc – you will make very slow progress. Goats vary a lot in temperament, and we have found that we get on much better with some than others. It is wise to start breeding your goats from a family which suits you. If you want your goats to produce household or commercial milk those with mismarks or wrong colours are just as useful as those which are correctly marked. In fact one well known breeder used to say that no good animal was a bad colour.

If we are going to improve goats we simply must get back to destroying our culls and stop selling them. I think that the general standard of present day show goats is a good deal lower than it was fifteen years ago and I am sure that this is due to lack of proper culling on the part of breeders.

Having decided that you will rear the kid you must see if it is horned or hornless. Figure 22 gives a rough guide to the way in which the hair grows on the heads of horned and hornless kids. You should be able to feel a small pimple at the base of the curl on the horned kid. If you are not sure about this clip off the hair round the site and see if there is also a little bare patch.

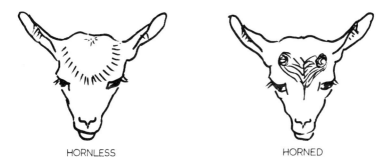

HORNLESS HORNED

22 How to tell the difference between a horned and hornless kid by hair growth

Your male kids *must* be disbudded at four days old. The horn growth on male kids is so strong that if you leave it later than this it is unlikely that the operation will be a complete success and there is nothing worse than deformed horns. Any kid with positive horns at birth should be disbudded at four days. For further information on disbudding see Chapter 11.

There are a variety of ways in which you can rear kids, the easiest being the natural way, leaving the kid on its dam. Unfortunately there are snags with this method. To start with the dam prefers to keep her milk for the kid and so when you come to milk her she 'holds back' her milk. She may do this whether the kid wants the milk or not, and if the milk remains in the udder it will start a process of reabsorption which gradually reduces the milk yield. When you finally take the kid away your milker will be yielding far less than she might have been had you been milking her and bottle feeding the kid. The other snag is that the kid grows up independent of you and when you want to train it to a lead or to parade for showing you will have a problem on your hands!

Bottle feeding a kid is a time-consuming occupation at first but by July the feeds should have been reduced to two a day which can be given at milking times. If, when the kid is born, it is put in a box or a separate compartment in the kidding pen, which can be reached by the mother's head, she can lick her kids and you can bottle feed them. At four days, when she goes out with the herd, they can go too and she will look after them. They will not try to suckle from her, but will look to you for their feed.

If you decide to take the kids right away from the herd for the first three or four weeks you will have to run them in separate enclosures with shelters and you will have to make sure that they start to eat as the adults do. Both these ways have their advantages but they do produce kids which are easily trained and, oddly enough, which grow to a greater size than those which are reared on their dams.

Bottles and teats must be kept as clean as all your other dairy utensils, otherwise the kids will get digestive upsets or refuse to take the bottle.

It is a waste of milk to feed a kid from a bucket or a pan, because when the milk is gulped down half of it fails to be mixed with the digestive juices in the mouth. It also causes the kids to get pot-bellied and to lack bone growth due to undernourishment.

When a kid is to be bottle fed it helps to give it the very first feed from a bottle, after which it can be taught to suckle its dam if you are leaving it on her for the first four days. When you take the kid away it will not be upset by the rubbery smell of the teat and you will have little trouble in teaching it to suck from the teat. A helpful tip when teaching a kid to suck is to put your hand lightly over the kid's head. (See photograph on page 54). Your hand simulates the goat's udder. If the kid refuses to take the teat you apply very gentle pressure with the finger and thumb at the corner of its mouth, and as the mouth opens you pop in the teat and close the mouth on it. Patience and time win round almost all reluctant feeders. Whatever you do to a kid, remember that it hates very firm restriction; the harder you hold it the more violent its reactions, so forethought and low cunning are your best weapons. The photograph on page 54 shows how to hold the teat.

For the first four days it is a good plan to feed five feeds of about four to five ounces of milk. If the kid is hungry and will take more, increase the amounts and knock off one feed. Three pints of milk in four feeds is quite adequate for a female kid; a male kid needs more. At about ten weeks, when the kid will take a pint at a feed, reduce the feeds to three a day. In July the midday feed can be dropped and in September you can reduce to one pint a day. Continue all winter if you have the milk, as it keeps the kid growing well. Minerals can be fed from the start in the kid's bottles, starting with a few grains and increasing to a pinch by the time the kid is twelve months old.

At a week you can tie up a stick of kale in the kid's pen and she will start to nibble at it. A small bunch of hay can be hung up too, but take care that there are no loops in which the kid can get entangled.

You should try to get kids to eat concentrates from an early date. Flaked maize (or for a first feed try a few cornflakes) is a good starter. They will only eat one or two flakes. Then try them with crushed oats. At this stage there is no need to worry about proteins since they get enough in their milk and from good grass pasture.

They should be encouraged to graze from two weeks onwards either with the herd or in their own worm-free paddocks. If yarded or kept in flat fields it helps them to develop if there is something they can climb on and jump off, and it keeps them out of mischief.

Kids' feet need trimming monthly; they also need regular worming, and they need a routine *Covexine* injection at five weeks. (See Chapter 4.)

Occasionally you may want to rear a kid when you have no milk. We have found *Vitamealo calf starter* excellent. You whisk two tablespoonfuls in a pint of warm water and it is ready to feed. The snag with this is that it is only sold in half-hundredweight bags, which is too much to purchase at one time if you are using it in such small quantities. The answer, if

23 Good way to carry a kid

you are lucky, is to find the local supplier and ask him which of his customers in your area uses it. Then you pay the customer a visit and hope he will sell you a small bag. It is important when changing over to *Vitamealo* that you substitute the powdered milk gradually, taking a week over the complete change.

Lamb rearing

Lambs can be most successfully reared on goats' milk, but to do it economically you must have grazing for them and be able to move them from one enclosure to another every seven days.

Lambs are best purchased directly from the farm and ideally should come from a multiple litter which the ewe cannot feed properly. Farmers are apt to leave the lambs on the ewe until one of them gets weak, and this makes rearing more difficult. If you know your farmer you should be able to purchase a strong healthy lamb. It is certainly easier to learn on this type than on a weakling.

The first feed should be three parts milk and one part water with one teaspoonful of honey. They will take up to 8oz of this mixture at a feed and for the first few days should be fed from a bottle four times daily. We keep our kids indoors until they are sucking their bottles well and they go out to pasture during the day. We provide them with shelter and hay at all times, and until they are three months old we bring them in at night.

Never feed lambs more than three pints of milk a day each. After about a week they will eat hay and this helps to speed up their growth. We also feed a small amount of flaked maize and crushed oats when they will take them, but with present prices I think this year's lambs will have to skip the maize flakes. We feed milk right up to the time they go to slaughter as they are there to use up summer surplus milk.

If you are selling them they should be ready when they weigh 80–90lb live weight, but if you want the maximum amount of meat for your freezer, 90–100lb is a better live weight.

If a lamb persistently scours after you have had it for several days, the best way of dealing with it is to get something from your vet as there are some very good remedies now. But honey is a splendid thing for lambs and once you get them going they should be alright. You need to worm them at six weeks and about every four weeks after that.

Calf rearing

Calves need at least a gallon of milk daily and unless you have a big surplus they are not an economical proposition. Some people rear replacement calves for a local farmer in exchange for hay or straw. Calves reared for veal are seldom an economical proposition. Calves reared on for 'baby beef' are, but you need a very strong shed, plenty of good pasture to move them round, plenty of hay, straw and concentrates, and they do very well with the addition of mangolds in the winter. When we rear either lambs or calves we keep them on milk as long as we can.

Calves will eat leftover concentrates, roots and pulled-through hay!

If you have kept your calf growing well and never allowed it to lose weight at any time it should be ready for slaughter at from fifteen to seventeen months.

It cannot be stressed too often that all utensils used for feeding milk to humans or animals must be washed and sterilised after use. Scouring (diarrhoea) in all young milk-fed animals is very often due to dirty utensils or milk which is fed too warm. The correct temperature is 103°F to 105°F. An animal's bodily temperature is higher than that of human beings.

10

SICKNESS AND HEALTH

There is no reason to suppose that goats are unhealthy unless they are badly managed or you are particularly unfortunate. Provided you worm them regularly, inoculate against clostridial organisms, keep their feet trimmed, delouse them during the winter, give them a couple of good shampoos in the summer and at all times keep them rotating on their pastures, there is every hope that they will be healthy.

Cold and hot weather present problems sometimes. Keeping the deep litter going in very cold weather helps to warm up the goathouse and should be sufficient in most thick-walled and well roofed buildings. I have a great desire to house our goats in a cavity-walled sunny house with the hay and straw loft above. This would keep it cool in summer and warm in winter, but alas, it is only a pipe dream! Thin-walled houses are a problem in cold weather. If the goats' hair stands on end making their heads appear much larger than usual, and if they shiver, you must do something to rectify the matter, provided there is nothing else wrong with them. If possible get them out and make them take some energetic exercise. It is good for you too, even if the snow is deep! If when you bring them in they are still cold you will probably have to resort to rugging them. This is never a very good idea, since if the cold spell lasts a long time it will be difficult to get the rugs off as their coats will sleek down. On the other hand, if you let them stay cold most of the food you give them will be used to provide body heat and the milk yields will fall. Rugging is cheaper; extra feeding is much easier but more expensive! Remember that once the yield drops in the winter it is very difficult to get it up again until the warmer weather comes.

Heat-waves present problems especially with high yielders. It has been found that if you feed ruminants quantities of green feed in hot weather the energy released digesting it adds further warmth to the animals' already overheated systems, causing much more discomfort. In these circumstances more easily digested concentrates are the answer. Another interesting discovery has been that if an ordinary wire-netting fence is erected with enough roof over it to cast a shadow, it causes a circulation of air which helps to keep the heat down a little. We have so few heat-waves in this country that this sort of thing is unnecessary, but it might give someone an idea for rigging up a temporary roof over an existing fence.

If your milker becomes so overheated that she pants with her mouth wide open it helps to cool her down if you wet her head, neck and part of her back with a sponge. This sort of thing sometimes occurs when goats entered in milking trials are housed in tents which become overheated when the sun is hot.

Flies are always a cause of much discomfort. There are several excellent cattle sprays on the market, some packed in aerosols, and it is quite easy to spray along the back of each animal as it is let out to pasture. It will give it some protection, and for the rest of the time available shelter is the answer. Flies dislike darkness and in hot spells and fly-ridden weather it is a good scheme to have blinds of some sort in the goathouse, provided they don't block out the air.

NURSING

However quickly you call your vet to a sick animal and however good his treatment, it will all be wasted if you do not nurse your goat properly.

She must be loose in a pen by herself, on a thick bed of straw. The pen must be light, free of draughts but well ventilated. If the goat has to be kept warm, rug her. She must be visited frequently. If she is lying flat out she needs to be propped up with straw. Sometimes if a goat has had a difficult kidding it is wiser to leave the kids with her. She will be more contented with them around and will make an effort to feed them.

Goats are very bad patients and seem to give up very easily, so every effort must be made to get them eating again. They can be given drenches made with milk and glucose and even an egg. A good mixture to give a goat daily after kidding, or if she has metritis or generally wants a build up, is made on a basis of milk to which two tablespoonfuls of brewers' yeast, a teaspoonful of minerals, two tablespoonfuls of *Cytacon* and a liberal helping of glucose are added. If you are lucky she might drink it, or take it through a teat, otherwise you will have to administer it as a drench. (See page 68.) Stout is another tonic.

The best cure of all is to get the goat to eat. Try to tempt her with some of the following delicacies: ivy; smooth holly; willow bark and twigs; raspberry leaves; brambles; thistles; hazel catkins and twigs; grass; red clover; goosegrass; campions; oats; vetches; ryegrass. Try apple peelings; carrots; baked bread rusks; baked potato peelings; pear peelings; pea pods; carrot tops; cauliflower leaves; purple sprouting broccoli; runner beans; French beans and parsley – all things which might tempt her. But you must go on trying.

A sick goat should always be given the best hay you can find, and should be placed where she can reach it. There are odd occasions when a goat has gone off her concentrates and is generally run down and you want to get her to eat something. Try a few *whole* oats. Sometimes this has an unexplained but beneficial effect on her and she starts eating again properly.

The main thing to remember is, when mixing up drenches which are to be given by mouth, not to give anything which will upset the flora in the rumen. If our goats have had antibiotics at any time we give them *Cytacon* and brewer's yeast to help re-establish the flora in the rumen.

If your goat has been very ill and, on visiting her you find her fast asleep, don't wake her even if it is time to give her a dose of medicine. Sleep is the best healer of all.

If the goat is unable to get up, you must try to get her up to urinate, and it is essential to turn her from side to side every four hours. Put her fodder where she can reach it.

When a goat has been very ill and has run a high temperature, when you go to trim her hooves later on you will find a thick ridge growing round the horn of her hooves, in much the same way as a ridge grows across a fingernail when it is going to come off. It takes some weeks for this to grow down, but you eventually cut it off when doing the routine hoof paring.

It might be as well to remind readers here that the Agriculture (Miscellaneous Provisions) Act 1968 requires the owner of farm livestock (including goats) not to cause unnecessary pain or distress to animals which he owns or to animals on land he owns. It also gives ministry officials the right to enter all agricultural buildings and land (except the farmhouse) and the right to insist that you have adequate buildings, light, heat, ventilation, drainage and water. The ministry has the right to insist that animals have a balanced diet. Other powers relate to control slaughter, restriction or interfering with the sight, hearing, smelling, or the emitting of sound. They also have prepared a code of practice for the care of all forms of livestock except goats, and the British Goat Society has produced that. This code is a list of minimum standards and was produced really for the use of the police and humane societies when they have to deal with cases of cruelty caused by neglect. (*See Appendix A*.) It is surprising how many goats are still kept in substandard conditions, in spite of all these precautions.

First aid and veterinary cupboard

Acetest pills
Anti-inflammatory and antibiotic cream (burns and scalds)
Antiseptic and antiseptic powder puffer (*Negasunt*)
Anti-spasmodic drench
Bloat drench
Cetavlon or *Cetramide* udder cream
Chloramphenicol eye ointment
Glucose
Hydrogen peroxide
Liquid paraffin
Obstetric cream or jelly
Oxytetracycline aerosol or 10% solution of formalin
Permanganate of potash
Shampoo (good cattle-type)
Stomach powder (Veterinary Drug Co Ltd, New York)
Vermifuges, as recommended by your veterinary surgeon
Cotton wool, crêpe and adhesive crêpe bandage, drenching bottle (narrow-necked), enema, foot-rot shears, hypodermic syringe (re-usable), lint, surgical scissors, thermometer, tweezers, *British Poisonous Plants*, Ministry of Agriculture & Fisheries Bulletin no 161 (HMSO)

Temperature, pulse and respiration The normal body temperature of goats is between 102.5°F and 103°F. When a goat has a fever its head looks larger than normal because the hair is erect, and the ears feel very cold. To take the temperature, wet the thermometer and gently slide it into the rectum.

The pulse rate should be from 70 to 80 beats per minute, and may be as much as 95 per minute in kids. The normal rate of respiration is about 20 to 24 per minute.

The new waste-food regulations

The Ministry of Agriculture has issued a circular which explains the new regulations governing the use of waste food:

Waste food – including swill – can be a very important factor in the spread of animal and poultry diseases and it is essential that every possible precaution is taken to prevent this risk.

If you collect or use waste food as a feedingstuff for livestock or poultry you will therefore be affected by the regulations contained in the Diseases of Animals (Waste Food) Order 1973.

The order defines livestock as: cattle, sheep, pigs and goats, and poultry as: all species of fowls, turkeys, geese, ducks, guinea fowls, pigeons, pheasants, partridges and quails.

The premises on which you process the waste food before feeding it to your livestock and poultry will be known as 'processing premises' *and must be licensed under the Order*.

Waste food is defined as: any meat, bones, blood, offal or other part of a carcase of any cattle, sheep, pigs, goats or poultry, or any product derived from such a carcase and any hatchery waste, eggs or eggshells.

Any broken or waste foodstuffs (including table or kitchen scraps) which contain or have been in contact with any part of a carcase are also included in this definition.

Bakery waste is regarded as waste food if it contains or has been in direct contact with meat in any form, including sausages, liver sausage, black pudding.

Meal manufactured from livestock or poultry protein is not included.

11

GOAT AILMENTS

R. C. Piggott

Chest disorders

COUGHS

A dry cough which affects all or most of the herd can be caused by feeding musty hay or using musty bedding (the fungus which causes this mould to grow will also cause infertility in your stock.) Neglected, this cough may turn to bronchitis and pneumonia. Humans can be affected and may develop bronchitis or farmer's lung.

Treatment Burn all musty hay and straw and replace with dust-free material.

PARASITIC BRONCHITIS OR HUSK

Symptoms A deep persistent cough.

Cause A worm (nematode) which migrates from the intestines to the lungs. Neglect will result in poor appetite, loss of milk and condition, and pneumonia.

113

Treatment   Call your veterinary surgeon.

PNEUMONIA
Symptoms   Coughing, wheezing and a temperature.
Causes   Neglected bronchitis, husk, damp buildings, damp bedding, transporting goats in cold, wet weather in open trucks, or excessive dust in very dry weather. Pneumonia can also be caused by stress and here the germ responsible is *Pasteurella haemolytica.*
Treatment   Call your veterinary surgeon at once. Isolate the goat in a dry, well-strawed pen and rug her.

MECHANICAL PNEUMONIA – see Choke in First Aid and Accidents (page 117).

Digestive upsets

ABDOMINAL ULCERS
Difficult to diagnose unless so bad they become perforated, when the goat may die, either from loss of blood or peritonitis. Goats with small ulcers periodically go off their feed.
Cause   Incorrect diet with too much concentrate feed and too little bulk feed leading to acidosis and a lack of muscular activity in the rumen.

BLOAT, BLOWN, HOVEN or FLATULENCE
Symptoms   This is acute indigestion. The rumen fills with gas and froth which the goat cannot expel. The left flank inflates and becomes hard. It presses on the diaphragm, making breathing very difficult. The goat is in acute discomfort, repeatedly lying down and throwing itself about.
Cause   The rapid eating of excessive amounts of succulent foods (particularly from spring pastures which have been treated with a nitrogenous fertiliser), lush clovers, vetches or lucerne, or gorging stolen concentrates.
Treatment   In mild cases, walking the goat around the yard

114

and then standing her front feet on a box and massaging the swollen flank should make her belch and relieve the situation. Use a bloat drench. If there is no sign of improvement you must call your veterinary surgeon. If left, increasing pressure on the diaphragm not only inhibits breathing but can stop the heart from beating.

COLIC

Symptoms   Acute pain in the abdomen. The animal attempts to lie in an unusual position, groans, rolls and attempts to kick its belly. Unlike horses, ruminants do not sweat when suffering from colic spasms.

Causes   Excess feeds of lush greens, especially if new to the goat, or frosted greens; or a foreign body (sand or grit, etc) or a poison. In kids it can be caused by violent exercise after a large feed of milk, or by constipation.

Treatment   Consult your veterinary surgeon. If he is not available, give about 8fl oz of liquid paraffin or the prescribed dose of an anti-spasmodic drench. It sometimes helps to give the goat an enema and this is certainly necessary with constipated kids. A dessertspoon of stomach powder (or as directed) in a tablespoon of liquid paraffin often helps (kids). We have found that once the goat dungs and urinates naturally the worst of the problem is over, but since we have taken more care in introducing new green foods or new pastures gradually we have not been bothered by this trouble.

CONSTIPATION

This is caused by lack of muscular movement in the bowels. It happens when poisons which act on the nervous system paralyse the nerves in the intestinal walls, which then absorb too much liquid; or when the animal has had too little to drink.

Treatment   Liquid paraffin should be administered by mouth, *as cold as possible*. This is not absorbed by the bowel and will protect it from further irritation. No other purgatives should be administered in cases of poison. The goat then needs a laxative diet with plenty to drink, molassed water, oatmeal gruel, syrup, yeast or warm milk.

GASTRO-ENTERITIS or PARASITIC ENTERITIS
Symptoms   Loss of bodily condition, anaemia and scouring, usually affecting several goats.
Cause   Generally due to heavy infestation of worms. If worms are *not* the cause, it is probably due to lack of proper nourishment and expert advice must be sought.

JOHNE'S DISEASE – see Infectious Diseases, page 127.

LUMPY DROPPINGS
These are signs of indigestion and worms.

RUMEN IMPACTION or CHRONIC INDIGESTION
Symptoms   A large, solid and distended abdomen which shows little sign of that muscular movement associated with rumination.
Cause   It sometimes occurs when goats are fed a large amount of concentrates and are then turned out to graze or browse. It can occur when poor-quality hay is fed with a concentrate feed of high fibre content. Sometimes it happens when goats are fed horse-nuts and poor-quality hay, because the horse-nuts contain a much higher percentage of molasses and this can prevent good digestion in ruminants. It is common among dairy goats fed on a low, inadequate protein and energy supplement when they are on a high-roughage diet.
Treatment   3 tablespoons stomach powder and 1 teaspoon salt in sufficient water to make 1 pint. Give as a drench three times daily until the trouble clears.

SCOURING or DIARRHOEA – see Gastro-enteritis and Johne's Disease (above), First Aid and Accidents (Poisoning, page 118), Infectious Diseases (Enterotoxaemia, page 125), Kid Ailments (Diarrhoea, page 128), Worm Problems (page 146).

Goat Ailments
First aid and accidents

BROKEN HORN
Call the veterinary surgeon.

BURNS AND SCALDS
Bathe in cold water. Dry and apply anti-inflammatory cream.

CHOKE
Caused by something obstructing the windpipe. Send for the veterinary surgeon. Choke can be caused by drenching if the goat's head is held too high and the liquid poured in rapidly. If fluid is thus accidentally forced into the lungs, even in small quantities, it can cause pneumonia; in larger amounts it is fatal – the animal dies from drowning. If this is the trouble, place the goat on a slope (board, built-up straw, anything for speed). With its hindquarters higher than its head, apply artificial respiration in much the same way as is done to revive anyone who has almost drowned.

COLLAPSES (from unknown causes)
The goat must be propped in a natural position, with bales of straw, etc, and its head should not be allowed to flop down. Cover its body with a rug or straw until the veterinary surgeon arrives.

CUTS, TEARS AND THORNS
Wash the wound, and if not very deep, puff well with *Negasunt*. For larger cuts, treat as above; then trim away the hair and try to stick the edges together with an adhesive plaster.

Deep holes made by nails or thorns, etc, should be cleaned well and bathed in hydrogen peroxide. If bleeding badly consult your veterinary surgeon.

ELECTROCUTION or STRIKE BY LIGHTNING
Both are usually fatal. Be sure, when wiring your goathouse,

that wires are put in conduits if there is any likelihood of the goats reaching them. Put all switches out of reach.

FRACTURES
Immobilise the limb and keep cold compresses on it until the veterinary surgeon arrives.

HAEMORRHAGE – see Torn Artery below

POISONING (CHEMICAL)
Causes   Weed killers, chemical sprays, air pollution near factories; all can cause poisoning if goats eat herbage contaminated by them.
Treatment   Call your veterinary surgeon and try to identify the poison.

POISONING (NATURAL)
Cause   Poisonous plants. Goats confined to browsing in woodlands without any supplementary feeding are liable to eat harmful plants which normally they would not touch. Housed goats can accidentally escape, and plants which free-range goats eat without harm can poison them. In periods of .drought, safe pasture-plants alter their chemical make-up and can become toxic: if these are the only source of nourishment they will cause trouble. Plants which become deficient in magnesium contain toxic levels of nitrates. This occurs in moist, warm, fast-growing weather, especially after the ground has been dressed with artificials, usually in spring and autumn. In the best regulated herds accidents happen, and goats get out and eat really poisonous plants. A regular daily feed of good hay and concentrates helps to offset these problems.
Treatment   *British Poisonous Plants*, Ministry of Agriculture & Fisheries Bulletin 161 (HMSO) was written by a man who kept goats. It tells you how to identify the plants, what effect they have on the goat and what to do by way of treatment.

Goatkeepers who take browsings to feed in hayracks should study this book carefully since animals in confinement will eat plants they would never touch when on free range.

SNAKE BITE
Most unlikely, but if it happens make the bite bleed and then apply permanganate of potash.

STINGS
Normally the odd bee or wasp sting will cause only minor discomfort but occasionally goats are stung by a swarm of bees or tread in a wasps' nest. Contact your veterinary surgeon at once.

SUDDEN DEATH
In all cases of sudden death you must call your veterinary surgeon.

TORN ARTERY
Symptoms   Blood pulsates from the wound in time with the goat's heartbeats.
Treatment   Call the veterinary surgeon at once. Apply a tourniquet above the wound, if on a limb, or grip the wound between finger and thumb using them like artery forceps. If blood is coming from deep in the wound apply real pressure with a dry lint pad and do *not* keep changing the pad or looking to see if the bleeding has stopped. Continue to apply pressure until the veterinary surgeon arrives.

Foot troubles

FEVER IN THE FEET, LAMINITIS, FOUNDER
Symptoms   The feet become hot and very tender, the horn becomes very hard indeed and the goat reluctant to walk, preferring to move around on its knees. The feet grow abnormally quickly and unless trimmed frequently become deformed. It

occurs more frequently in the front feet, but can afflict the hind feet as well. The goat is reluctant to lie down at first and paddles from one foot to the other. Later it may lie down and refuse to get up.

Causes A bad kidding, metritis, pneumonia, mastitis, a very sudden change in feed, or an allergy. It can follow one of the clostridial infections. Chronic laminitis is most often the result of over-feeding concentrates, particularly proteins.

Treatment Locate the cause with the help of your veterinary surgeon and treat that. Soak the feet daily in warm water. Trim them at least twice a week and spray with chloramphenol spray. If the cause was a sudden change in feed, a dose of 8oz milk of magnesia in 4 to 6oz liquid paraffin will help to clear any dietary upsets. 'This complaint is the bane of herds bred for high yields when they are inexpertly managed,' comments S.B. Guss. We are lucky in this country in that we have available an abundance of green fodder most of the year.

Nursing a goat with laminitis, apart from the treatment, is time-consuming. She needs brisk exercise about four times a day in order to keep her digestive organs working normally. She must have clean dry bedding at all times and her food, water and mineral lick must be placed where she can easily reach them.

FOOTROT

True footrot (see also Mud Fever below) is caused by a bacterium, *Fusiformis nodosus*, which is carried in the foot and which can remain active in pasture for fourteen days. It starts as an inflammation between the claws of the foot and spreads under the horn of the hoof causing the horn to separate from the fleshy part. If left it will spread to the whole foot. It causes pain in varying degrees and the goat becomes lame. If neglected the animal is unable to stand on the affected feet.

Treatment The hoof must be pared back (Fig 15), and where the horn has separated from the side it must be clipped away. The whole foot should then be sprayed with a chloramphenol

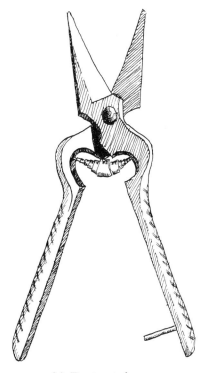

24 Foot-rot shears

spray, or the foot can be soaked in a 10 per cent solution of formalin.

Prevention The infected pasture must be kept clear of animals for two weeks, when after a further treatment the goats can return. The pasture they were on whilst under treatment must be rested for a further two weeks.

LAMENESS

Causes Injury, rheumatism, footrot, mud fever, or a stick or stone jammed between the claws. Rheumatism is caused by the goats having to live on wet bedding. They have difficulty in getting up, and walk stiffly with their backs arched.

121

Treatment   Consult your veterinary surgeon.

MUD FEVER

This is not infectious but looks the same as footrot. It is caused by *neglect*. Overgrown horn collects pockets of dung and mud which forces the horn away from the side of the foot.
Treatment   As for footrot.
Prevention   Cut your goats' feet monthly.

OVERGROWN HORN – see Chapter 8 and Fig 15.

STRAWBERRY FOOTROT

Symptoms   The skin of the foot and up to the hock or knee becomes moist and inflamed. This dries and small scabs form, and when these drop off they leave small ulcers. These can spread to the teats of a badly affected goat.
Cause   A fungus which usually appears during a warm, wet spell of weather.
Treatment   Consult your veterinary surgeon.
Prevention   A good regular hoof-paring programme will prevent most outbreaks.

Head troubles

BLINDNESS

Can be caused by injury, or the effect of lead or other poisoning: See First Aid and Accidents (Poisoning, page 118), and Pink Eye (below).

CYSTS IN TASSELS – see Kid Ailments, page 128

DEHORNING

Horned goats should not be kept in a milking herd and it is illegal to keep them in pens with hornless goats or to put them together during transport.

But dehorning an adult goat is not a job to be undertaken

lightly. The goat should be 2 years old or more, when horn growth has slowed down. The operation must be done by a veterinary surgeon under a general anaesthetic. A bandage, plaster or hood must be kept on the wound for at least a week after to prevent dust, chaff or hayseeds entering the sinus. It must not be performed at the time of year when the fly problem is bad, nor must it take place in cold frosty weather.

The obvious answer is never to buy a horned goat and to make sure all your kids are disbudded.

DISBUDDING – see Kid Ailments, page 129

ORF or POSTULAR DERMATITIS - see Skin Diseases, page 143

PINK EYE or CONTAGIOUS OPHTHALMIA
Symptoms   It starts with a running eye and this is the time when treatment is most effective. If left, the goat will be in considerable pain, the eye clouds over and pus exudes from it. Ulcers form and it can lead to total blindness. Treatment is then very lengthy and the goat has to be kept in isolation for as long as the eye has to be dressed.
Cause   An organism called *Rickettsia conjunctivae* which invades when there has been accidental damage to the eye. The trouble is most often seen in dry, dusty and windy weather when the eye may get scratched. It is highly infectious.
Treatment   Apply eye ointment daily, as advised by your veterinary surgeon, and keep the goat isolated until cured.
Prevention   It is wiser to treat all weeping eyes with chloramphenicol eye ointment.

THYROID GLAND, ENLARGED
Symptoms   The thyroid gland is situated in the throat. Due to hereditary factors and a lack of iodine, some kids are born with an enlarged gland which feels soft when handled. Kids with

this problem often feed well and grow quickly, but put on excess fat unless given supplementary iodine.

Cause   Lack of iodine. Some areas are deficient in iodine. Some plants, wherever they are grown, inhibit thyroxine production – cabbages, soya beans and yellow turnips. During the processing of soya bean meal, the goitrogenic substance is partially destroyed.

Treatment   Consult your veterinary surgeon and he will supply you with iodine drops to put in the kid's bottle.

Prevention   Any mineral salts fed to goats with concentrates should contain 0.007% iodine. In addition, an iodised mineral lick should be available in the stall so that any inadequacies can be corrected by the goat itself. Kids can be fed this mineral once a day in their bottle (see maker's instructions for the correct amount). Some people feed kelp, an organic source of iodine, in the concentrates, but this is not regarded by the veterinary profession as being very helpful.

Infectious diseases

ABSCESSES

These can appear on any part of the body, mostly as the result of an infected wound.

Cause   Abscesses which form in the udder are often the result of pasteurella, streptococci or staphylococci infections in mastitis, etc. Such infectious abscesses can lead to very serious troubles. Lumps which form at the site of a vaccination injection are not infectious abscesses and if left alone will eventually burst and slough off. They are caused by the goat having a reaction to the fluid which carries the vaccine.

Treatment   Consult your veterinary surgeon.

Prevention   Dirty bedding, dirty stalls, overstocked and infected pastures and generally insanitary living conditions are the main sources of infection. If you let your goats lie on wooden benches and slatted floors it is almost impossible to keep them clean.

ANTHRAX

This is a *notifiable* disease and if not reported promptly to your veterinary surgeon you are liable to heavy fines.

Symptoms   Mostly sudden death is the only indication of the disease. If observed in the initial stages the goat will have a high temperature (107°F to 108°F), it will be severely depressed, have difficulty in breathing and great weakness.

Cause   A germ called *Bacillus anthracis*, which can live for many years in the soil or can be transmitted in imported feedstuffs, eg bonemeal.

BLACKLEG – see Enterotoxaemia below.

BRUCELLOSIS or CONTAGIOUS ABORTION

Cause   *Brucella abortus*. The Ministry is in the process of eradicating this disease from cattle. It causes undulent fever in humans, which is a serious illness. They are quite satisfied that it does not occur in this country amongst goats and they do not require goats to be tested when cattle on the same farm are having routine tests.

Brucellosis melitensis *does* affect goats. Commonly known as Malta fever, the last case known in this country was in 1922 in a goat brought back from Malta.

ENTEROTOXAEMIA

Symptoms   Mostly sudden death; or wobbly gait, collapse and scouring together with a peculiar call, unlike most other sounds the goat makes, followed by death.

Cause   *Clostridium welchii*, type C, which is found in the digestive tract. It is not fully understood why this and allied organisms, also present in the environment, suddenly flare up and produce deadly poisonous toxins.

Treatment   None.

Prevention   There is no need to expose any goat to the risk of any of the clostridial diseases as there is an excellent programme of subcutaneous injections which will give the herd immunity. Consult your veterinary surgeon about this and get him to teach you how to do these injections. A 50ml pack of *Covexine* will do the two initial vaccinations for seven goats and after that they will need two 2cc injections twice yearly. The 'shelf life' of these vaccines is very short, so that once the initial vaccinations have been done, goatkeepers who only have two or three animals should purchase individual doses from the veterinary surgeon.

The types of clostridial infection covered by this vaccination are as follows:

Cl welchii type D	Lamb dysentery
Cl welchii type C	Struck or enterotoxaemia
Cl welchii type D	Pulpy kidney enterotoxaemia
Cl septicum	Braxy
Cl chauvoei	Blackleg or post-parturient gangrene
Cl oedematiens type D and *Cl tetani*	Tetanus

Most of these diseases are fatal and all are infectious.

FOOT AND MOUTH DISEASE
Goats seem remarkably resistant to this disease but if there is an outbreak on a holding where goats are kept they have to be slaughtered with all other cloven-hoofed animals.
Symptoms   Dullness and lack of appetite, small blisters on the feet causing lameness and little pawing movements of the foot. Dribbling from the mouth, which is so noticeable in cattle, is much less marked in goats. Blisters appear in the mouth and on the tongue.
Treatment   Send for your veterinary surgeon at once. This is a highly infectious and *notifiable* disease.

FOOTROT – see Foot Trouble, page 120

GROWTHS
Occasionally these appear and should always be investigated by your veterinary surgeon.

JOHNE'S DISEASE
This is uncommon in goats but has been found in well-bred goats kept in small, overcrowded paddocks.
Symptoms   Persistent loss of condition, harsh coat and incurable scours, often intermittent. In goats, however, this is not a common symptom until a few days before death. There appears to be a greater risk of infection in kids under the age of 8 or 9 months. Older animals are resistant enough not to show the disease.
Diagnosis   Laboratory culture of dung samples.
Treatment   Must be left to your veterinary surgeon.

MALTA FEVER – see Brucellosis above, page 125

MASTITIS – see Milk Production Troubles, page 132

ORF or PUSTULAR DERMATITIS – see Skin Diseases, page 143

PINK EYE or CONTAGIOUS OPHTHALMIA – see Head Troubles, page 123

POX – see Skin Diseases, page 144

PULPY KIDNEY – see Enterotoxaemia above

RINGWORM – see Skin Diseases, page 145

STRAWBERRY FOOTROT – see Foot Troubles, page 122

TESTICULITIS – see Male Disorders, page 132

TETANUS – see Enterotoxaemia above

TUBERCULOSIS
This is relatively unknown in goats. It has been known in kids which have been reared on *infected* cows' milk.

Kid ailments

COLIC – see Digestive Upsets, page 115

CYSTS

Kids are sometimes born with a cyst in a tassel which may range in size from a small pea to much larger. Veterinary surgery is needed to remove them.

DIARRHOEA or SCOURS

Uncommon in kids suckling their dams, but probably due to a sudden change in the dam's diet or because she has eaten a poisonous plant. Treat the kid with a mild antacid, such as Macleans Stomach Powder. In bottle-fed kids scours is seen all too frequently.

Causes Milk-feed too warm, sudden change in diet, dirty bottles and teats, *E coli* infection, worms.

Treatment Milk must never be hotter than 102°F. Never suddenly change the diet of any ruminant. Rinse all bottles and teats *twice* in cold water as soon as the feeding is completed. Sterilise bottles and teats at least once daily in the same way as other dairy equipment.

E. COLI INFECTION

Symptoms Brownish-yellowish scours which become white and then watery. Temperature rises to 105°F at the onset of the disease and then falls rapidly as the disease progresses. The kid becomes weak and dehydrated. In acute cases death is fairly rapid.

Cause *E. coli* organisms build up in areas where the kids are born unless strict hygiene is observed. If a succession of goats kid in the same area the problem gets progressively worse. If the kid has no colostrum (because of pre-milking the dam, removal at birth or dam has no milk) it has no resistance to these organisms.

Treatment   If there is an *E. coli* problem in your herd you would be well advised to discuss the problem with your veterinary surgeon. New-born kids can be inoculated at birth and there are other forms of protection as well as following preventative measures.

Prevention   Goats should kid in a pen which has been thoroughly disinfected and which has a good bed of new dry straw. Do not use dry bedding from other pens. If kids are to be reared away from the adults it is best done in paddocks which have not had kids in them for at least twelve months. The kids should be moved to new paddocks every seven days.

WORM INFECTION

Symptoms   Dark, bloodstained scours.

Cause   Coccidia infection.

Treatment   Consult your veterinary surgeon.

Prevention   Clean pasture as in *E. coli* infection above. Worm the kids regularly (see Worm Problems, page 146).

DISBUDDING

It is illegal to disbud a kid without an anaesthetic after it is 2 weeks old, but it is much better for the kid and the operator if it is anaesthetised before this.

Kids should never be disbudded by a novice. Caustic sticks used carelessly can blind a kid, and should never be used on a male kid as they seem to have little effect on their positive horn growths. Overheating the brain with the hot iron will cause meningitis. A kid disbudded by an inexpert operator often suffers acute pain and for the rest of its life flinches away from contact with the top of its head. When disbudded by a veterinary surgeon kids seldom seem to notice the operation, but if they do grow scurs around the horn bud which cannot easily be removed, they must be taken back for touching up. Most veterinary surgeons include the charge for 'after care' in the initial fee.

HYPOGLYCEMIA or BIRTH CHILLING

Symptoms   Kids' coats stand on end, they shiver, arch their backs and finally – if left – lie down curled up, become comatose and die.

Causes   Kids born in very cold weather, very weak kids, kids whose dams have no milk or who refuse to let them suckle become rapidly depleted in glucose in the blood.

Treatment   Put the kids in a warm place (80° to 90°F) and dose by mouth with warm *Ovigest*. Either use a hypodermic syringe to dribble it in or use a lamb reviver (a plastic container with a tube which can be gently inserted down the gullet and the dose is then delivered into the stomach). When the kid has revived a little give it 2oz colostrum, by the same method if it will not suck a teat. As soon as it becomes normally active it can be returned to its dam or the rearing pen.

JOINT-ILL

Symptoms   Rising temperature (105° to 106°F) which causes the kid to stop feeding. It hunches its back, one or more joints swell, and it becomes lame.

Cause   Streptococci, *Corynibacterium ovis* and *Bacterium coli* present in dirty bedding or pasture at the time the kid is born, which gain access to the bloodstream through the navel cord.

Treatment   Call your veterinary surgeon at once.

Prevention   Provision of clean maternity pens (see *E. coli* Infection, page 128). Dip the navel cord of all new-born kids in iodine.

RICKETS

This is rare in young stock on pasture but occurs in stall-fed kids and goatlings as a result of mismanagement. A lack of good quality fresh grass and hay and a concentrate ration of bran and oats brings about an imbalance in calcium and phosphorus.

Treatment   Consult your veterinary surgeon who will advise on vitamin D injections and the inclusion of calcium and phosphorus in the diet. Let the kids have space to take plenty of exercise in the sunshine, feed adequate greenstuffs and better hay, and leave the bran out of the diet.

ORF or PUSTULAR DERMATITIS – See Skin Diseases, page 143

STRESSES
All kinds of stresses in animals lower their resistance to disease.
Environmental Kids can stand very cold temperatures provided they are fed adequately on milk or milk substitutes. They must have protection from wind, rain and draughts and they must have a clean dry bed free from moulds and damp.
Due to erratic feeding Kids should receive their bottle feeds at regular intervals to prevent them gorging themselves, otherwise they will suffer from enteric troubles.
Spatial Kids need adequate space when being reared in groups otherwise the dominant ones will bully the timid ones. They need plenty of exercise, preferably outside in sunny weather. When indoors, the pens should be light and airy with room for them to run about and objects for them to jump on and over.

THYROID GLAND, ENLARGED – see Head Troubles, page 123

Male disorders

HARD SCABBY SKIN
This condition sometimes occurs in male goats after the age of 3 years. If it is neglected it becomes chronic.
Treatment Wash with warm rainwater and soft soap, rubbing and massaging his skin until he is a mass of lather and his skin saturated. Rinse well and dry him. Before the pores close after this shampoo, massage with 5 per cent carbolised oil. Make sure he is not saturated with oil otherwise he could

catch a chill. This treatment should be repeated every few days until the skin softens and new hair begins to grow. It is best to do this in late spring when the weather is warm.

INFERTILITY

This is fairly common in male goats. It can be caused by unsuitable feeding, a blocked urinary tract, an inherited characteristic (cryptorchidism), or it can be temporary due to over-use or cystitis.

Male goats should not be fed barley, mangolds, swedes or sugar-beet and should be given rainwater to drink.

Blockages in the urinary tract are caused by calculi which make it difficult or impossible for the male to urinate. He may pass minute quantities of blood-stained urine. Send for the veterinary surgeon immediately.

Cryptorchidism is a condition in male kids where one or both testicles have failed to come down into the scrotum. These males will be infertile.

Traumatic Testiculitis   When male kids are reared together and are very active, one may receive a severe blow in the testicles from an aggressor. As soon as a male kid becomes aggressive he should be removed and penned on his own.

ODOURS – see Chapter 7

Milk production troubles

BLOOD IN MILK – see Clinical Mastitis, page 134

CLOTS IN MILK – see Clinical Mastitis, page 134

GOAT POX ON UDDER – see Skin Diseases, page 144

HYPOMAGNESEMIA – see Nutritional Deficiencies, page 136

MASTITIS

Symptoms   Inflammation of the mammary glands in the udder, blood or clots in milk.

132

Causes Blows, injuries or cuts, sore teats and spots all allow bacteria to invade and infect the udder. Lowered resistance due to some other illness will also allow infection to take hold. Infrequent milking resulting in the udder becoming hard and distended, or failure to milk when the udder has reached this stage either before or just after kidding will damage the udder-tissue and allow bacteria to infect it. If the goatkeeper is too rough when milking, dirty milking techniques, dirty housing and bedding, or making goats with large udders sleep on benches or slatted floors, are all situations favourable to bacterial udder infection. Failure to clean the udder prior to milking, to use an udder cream during milking and a teat dip after milking and to treat even small scratches promptly all can cause mastitis. Shows are a real source of infection since stripping stewards go from one herd to the next without washing their hands and in the show ring the judge simply hasn't time to wash between examining one goat and the next. The mildest form of mastitis – sub-clinical streptococcal (see below) – is acutely contagious and is very frequently spread this way.

Prevention Goats *must* be housed on dry, clean straw. Deep litter is excellent provided it is properly managed with a good layer of dry, clean straw on top. Horned and hornless goats must never be kept in the same pen or loose housing.

Spots, scratches, cuts and other injuries to the udder must be treated at once and never neglected. Sore teats, chaps and slight injuries to udder and teats must be treated with *Cetramide* udder cream. The milking routine must be strictly adhered to and never skimped (see Chapter 5).

In all types of mastitis, when the trouble has cleared, further laboratory tests should be carried out to ensure that all bugs have been destroyed.

SUB-CLINICAL MASTITIS

Symptoms There is frequently nothing to suggest that a goat is suffering from this, there are no clots in the milk and it often boils satisfactorily. Low butterfats and reduced milk yield suggest that milk samples should be taken and analysed, and if streptococcal infections are present your veterinary surgeon will advise you about treatment. Very many producers who advertise their milk as suitable for feeding to very young babies without pasteurising, are completely ignorant of the fact that it may be infected and could cause serious problems to the baby. All who produce milk for sale should have it tested regularly however tedious this may be.

CLINICAL MASTITIS
Symptoms Reduced milk yield, temperature often 104° to 105°F. There are swellings, tenderness and heat in the udder; the milk is watery with white flakes or larger clots and there may be some blood.
Cause *Streptococcus agalactiae.*
Treatment Call the veterinary surgeon at once. Neglect to treat any form of mastitis as soon as it starts can cause permanent damage to the udder and thus prevent the return to full milk production.

SUMMER MASTITIS
Symptoms Marked signs of illness, the udder is inflamed and painful and the secretion extracted from it has an offensive smell. The onset of the disease is *very* sudden, quickly followed by collapse and often death.
Cause A staphylococci infection. This is acute gangrenous mastitis.
Treatment Call your veterinary surgeon at once.

MILK FEVER
This occurs just before, during, or just after kidding.
Symptoms In severe cases the goat becomes unconscious, but in less severe cases she walks with a staggering gait, is often constipated and her hocks become very straight. There is an inability to bend or flex these joints. If it happens during kidding

134

she will lose those strong muscular contractions which expel the kids and the afterbirth. Her hindquarters feel much colder than her forequarters.

Cause Insufficient calcium. This can be the result of a poor diet or, surprisingly, too much calcium in the diet. It is thought that excessive amounts produce substances which block the easy passage of calcium into the bloodstream, so the goat has no reserves to call on when milk production increases.

Treatment Call the veterinary surgeon at once and he will administer an intravenous injection of calcium, etc. In the meantime the goat must be propped up in a natural position with her head pulled to one side to prevent her inhaling the cud which is being forced upwards by gas in the rumen.

It is now thought that stripping the udder out after kidding – or pre-milking – has little to do with the cause of milk fever. S.B. Guss suggests that it might be wiser not to completely strip out the udder for one day following the treatment.

PINK MILK

Freshly kidded, high-lactating goats sometimes produce pink milk or milk which, if left standing, produces a pink sediment at the bottom of the container. Sometimes there will be very small spots of blood on the filter pad after the milk has been strained.

Treatment The novice should consult the veterinary surgeon. Generally speaking, if the goat is well and her yield rising there is nothing to worry about, since it is probably a small blood vessel which has ruptured and the trouble will clear up naturally. Do not sell pink milk.

TEAT DUCT, CLOTS or GRANULES IN

These often cause trouble at the start of a heavy lactation, and you must gently manipulate them out of the teat. If the goat is eating well and her yield rising, it is usually a sign that she is going to produce a good lactation.

Goat Ailments

Nutritional deficiencies

ANAEMIA

Symptoms   Very pale lips and membranes inside eye-lids.
Causes include iron deficiency, which is common in animals
fed on an all-milk diet; but since kids browse, graze and eat hay
from a very early age this should not occur. It can also be
caused by a deficiency in cobalt and certain poisons such as
lead.

Another cause is an infestation of bloodsucking worms,
coccidia (see Worm Problems, page 146).
Treatment   Feed a good goat-mineral and always have
iodised mineral lick available in the stalls.

HYPOMAGNESEMIA or GRASS STAGGERS

Symptoms   Excitement, fits and sudden death if illness is
acute; nervousness and trembling especially when being
handled or milked if the trouble is chronic.
Cause   Lack of magnesium – lush grass in spring and autumn
prevents the proper absorption of this mineral. If soils deficient in magnesium are dressed with potassium, it makes this
condition more likely.
Treatment   Call your veterinary surgeon.
Prevention   Where lack of magnesium in the soil is known to
exist, it is better to graze for a month in the spring before dressing with potassium. The first growth can be cut for conserva-
tion if this is not possible. In all cases supplementary feeding
on hay and concentrates will help to prevent this trouble.

MOULD INFECTION

Avoid mouldy foods and bedding. They cause serious chest
disorders and infertility and can be fatal, the mould resulting
in a lack of thiamine in the rumen which prevents the production of vitamin B. Mouldy hay and straw will also affect the
goatkeeper and cause lung troubles including farmer's lung.

PINE
Symptoms   Anaemia, lack of appetite, poor milk production, a coarse dry coat.
Cause   Lack of the trace element cobalt.
Prevention   Make sure that the minerals you are using contain cobalt. Find out from your veterinary surgeon if you live in a cobalt-deficient area and ask his advice.

Reproductive disorders

ABORTION
Cause   Accident, fighting, intestinal upset or poison.
Treatment   In all cases consult your veterinary surgeon (see Brucellosis, page 125).

ACETONAEMIA – see Pregnancy Toxaemia below

ANOESTRUS (failure to come into season)
Cause   Anaemia (see Nutritional Deficiencies), malnutrition, or extreme cold. Sometimes in older goats it can be due to a heavy milk production.
Treatment   Consult your veterinary surgeon.

CYSTIC OVARIES
Symptom   The goat appears to be permanently in season.
Cause   A cystic condition which prevents the goat from ovulating at the proper time.
Treatment   Call your veterinary surgeon. Treatment needs to be successfully completed by January to allow time to get the goat mated. If it is unsuccessful, it would be wiser to cut your losses and have the goat destroyed.

FALSE PREGNANCY (CLOUDBURST)
Symptoms can occur both after a normal mating has taken

place or when a goat is being 'run through'. It sometimes happens in unmated goatlings. The goat's body increases in size as if she is in kid but after four months the skin round the vulva has *not* become puffy and swollen as it does when the goat is carrying kids. Another indication for the very observant is that at this time the 'in kid' goat places her feet on the ground with a very heavy tread because of the growing weight of her kids, etc, whilst the goat which is going to cloudburst is as merrily light-footed as all the other unmated goats. After about the normal gestation period she will produce 2 or 3 gallons of watery fluid. After a couple of days this clears up and she sometimes comes into milk or, if she is already in milk, the yield goes up.

Treatment The goat must be remated the first time she comes in season if not in milk, otherwise she can be safely left till the following autumn.

INFERTILITY (FEMALE)

If your goat regularly returns to service, suspect the male and try a different one. Much rarer is the goat who comes normally in season but refuses to stand for service by the male of your choice. Even if you succeed in getting her served, she returns after three or six weeks. Before you do anything else try using another male. Goats sometimes have their likes and dislikes!

If the female returns and on inspection has a dirty looking discharge she may have vulvitis, an infection of the vulva caused by flies attracted to it by damage to the lips from scratches or pustular dermatitis (see page 143) or she may have vaginitis after a difficult kidding from dirty hands assisting at kidding or following metritis (see below). It could mean that the goat is unwell. In all these cases, consult your veterinary surgeon.

The goat may be a hermaphrodite (see Chapter 9).

Failure of the goat to come into season by Christmas may be due to anoestrus (see above); or to cystic ovaries (also see above) which may be causing late ovulation. In this latter case if you have no reason to suspect the male you should seek veterinary advice, because a lutenising hormone is needed. The same applies to the goat who comes violently in season at the correct time or erratic times.

INFERTILITY (MALE) – see Male Disorders, page 132

INFLAMMATION OF THE BLADDER
Very unusual in goats, but sometimes occurs after a difficult kidding.
Symptoms The goat frequently passes small amounts of urine with some straining.
Treatment Consult your veterinary surgeon. Where the liquid stains the goat, small doses of bicarbonate of soda prevent scalding and loss of hair.

KETOSIS – see Pregnancy Toxaemia below

METRITIS
This is inflammation of the lining of the uterus.
Causes Infection by a retained afterbirth; infection introduced when assisting kidding; lack of vitamin E in areas where there is a selenium deficiency.

Acute Metritis
Symptoms High fever (temperature 105° to 107°F), severe depression, indifference to kids, no appetite, discharge of dark-red pus with a very unpleasant smell.
Treatment The veterinary surgeon must treat this. If neglected it can become chronic and ruin the lactation, and possibly make the goat sterile.

Chronic Metritis
Symptoms A mild fever, but apart from an abnormal and unpleasant discharge there are no other symptoms.
Treatment It may abate naturally. If the cervix is sufficiently dilated the veterinary surgeon can give the goat a wash-out. On no account should this be attempted by the amateur.

Prevention Keep the goat on clean straw and call the veterinary surgeon if the afterbirth is retained longer than 20 to 24 hours; he will inject an antibiotic. If assistance is needed at kidding time, always insert a suppository when you have finished; your veterinary surgeon will supply you with these. Make sure that selenium is included in the mineral salts you feed to your goats.

OBTURATOR PARALYSIS

Symptoms Partial or complete paralysis to one or both hind legs and the tail.

Cause Undue pressure and damage to the obturator nerve during the passage of the kid through the birth canal.

Treatment Time and nursing are the only answers to this problem. Always consult your veterinary surgeon. It is most important that a goat in such a collapsed condition is turned manually from one side to the other *every 4 hours*. She must be kept on a deep bed of dry straw and must be propped on her brisket. Food and water must be placed where she can reach it and, if the weather is cold, she needs to be covered with a rug. She must be milked regularly.

PREGNANCY TOXAEMIA, ACETONAEMIA AND KETOSIS

Symptoms Pregnancy toxaemia is characterised by a poor and finicky appetite. A sample of the urine should be tested. Sometimes the legs – one, some or all – swell. The whites of the eyes become dirty looking. The symptoms of acetonaemia and ketosis are the same as those of pregnancy toxaemia except that the legs do not swell. Diagnosis and treatment are the same.

Cause To quote S. B. Guss:

This occurs in dairy goats in two situations, as pregnancy toxemia within the month before kidding or as a ketosis following kidding. In the former the disease is more common in over-conditioned goats which have had little exercise. Accumulation of abdominal fat during late lactation and early pregnancy limits the amount of feed the goat can eat. When the demands of the rapidly growing kids occurs the body attempts to utilise stored body fat and ketosis occurs. It also occurs in underfed goats with multiple foetuses during the same period.

Treatment   Call your veterinary surgeon.
Prevention   We have found that the trouble clears very quickly if you catch it in time, so we test urine regularly during the last month of pregnancy and for about a month after kidding. If there is a reaction we give the following treatment using the propylene glycol *Ketol*.

Take 3 drenching bottles. In one put 2oz of *Ketol*; in the second 1 tablespoon bicarbonate of soda mixed with 4oz water; in the third 1 cup of glucose or honey mixed in water.

Dose with the bicarbonate first, then with the sugar and lastly with the *Ketol*. The bicarbonate closes the oesophageal groove, and the sugar and *Ketol* are delivered straight into the abomasum (true stomach) for immediate absorption. Repeat this twice daily for 2 days. Afterwards continue the glucose for a few more days if you think it is necessary.

PROLAPSE (OF CERVIX AND VAGINA)
Symptoms   This usually happens during the later stages of pregnancy. A protuberance, which may be either hard or soft and which looks red and raw, is seen at the vulva. The goat does not appear to bother about the condition.
Treatment   Call the veterinary surgeon who will deal with the protuberance by stitching it back into position. The goat will probably kid quite normally and often the stitches break without assistance from the goatkeeper. Since the condition is highly likely to recur it is probably wiser to have the goat destroyed at the end of the lactation.

PROLAPSE (OF UTERUS)
Symptoms   This is much more serious. It occurs after the kids have been born, within an hour or two. It is more usual in older

goats. The prolapse is far bigger, redder and covered with button-like protuberances. The uterus sometimes hangs right down to the ground. At first the goat will seem unconcerned but she soon becomes distressed, lies down and strains and then becomes very dejected.

Treatment Send for the veterinary surgeon at once. Keep the uterus warm by bathing it with a blood-heat (102°F) saline solution until he arrives. To quote the TV vet, Lois Hetherington, in *All About Goats*:

> Nowadays, because of the availability of antibiotics, the success rate of treatment is excellent. Not only does the mother recover completely, but she is usually able to breed again. It is my experience that female goats rarely prolapse the uterus twice.

RETAINED AFTERBIRTH

A goat should expel the afterbirth within an hour or so of kidding and certainly before 20 to 24 hours. On no account should any attempt be made to pull it away. If after this time it is still there, call your veterinary surgeon.

RICKETS IN PREGNANT GOATS

Symptoms It starts with stiffness in the legs, neck and back.

Cause This rarely happens, but has been known in goats which do not dry off during pregnancy.

Treatment Call the veterinary surgeon.

WHITES

This is an old-fashioned name for chronic catarrhal metritis (see above).

Skin diseases

ABSCESSES

Symptoms A rounded swelling probably with a healed scab or scar over it.

Cause Infection in a scratch, wound or open sore.

Treatment *Never* lance a swelling. Poultice it if it is causing discomfort, but leave it alone if it does not bother the goat. Always consult your veterinary surgeon about any lumps (see Infectious Diseases, page 124).

Prevention Keep your goats on clean dry bedding and dress scratches and wounds with *Negasunt* or *Cetramide*.

ORF or PUSTULAR DERMATITIS

Symptoms Small blisters round the mouth and nostrils which rupture and become scabby with inflamed skin under them. Secondary infections can invade these spots and make the trouble much worse. When badly infected a kid will refuse to feed, and if kids suckling their dams contract this disease they spread the infection to her teat canals, where it is very painful and often causes mastitis.

Cause An acutely contagious virus which is highly resistant to drying, and which has remained active in scabby material for as long as twelve years. Because of this a herd is liable to further outbreaks although the animals which have had the disease are probably immune. Orf is sometimes transmitted when animals browse on gorse which has been infected by other animals, the virus entering through the pricks made by the gorse.

Treatment The whole herd should be immunised when there are no longer kids suckling their dams. After this it is only necessary to vaccinate new kids when they have finished suckling their dams and before they go to their first show. All goats who go to shows should be vaccinated. Keep living quarters clean.

When dressing spots wear protective gloves; this virus can infect humans too. Spray with a purple aerosol containing oxytetracycline obtained from your veterinary surgeon and discuss an immunisation programme with him. The spray will cure human orf infections too!

143

PARASITES

Fleas and Lice

Symptoms Biting and bloodsucking insects live in the goat's hair making it dry and scurfy and causing irritation. The trouble is easily seen because whenever the goat scratches with her teeth, the hairs part giving her coat a gappy appearance. If the goat has been ill the parasite population increases rapidly.

Treatment Dust with a good cattle-louse powder every 10 days until the goat is clean.

Prevention Regular dusting during the winter months for routine control (see also page 90).

Ticks

These are mostly summer visitors. The insect lives in pastures and in bracken, attaches itself to any suitable place on the animal and sucks its blood. When full, the tick is a noticeable rounded ball. Remove carefully, holding it with your finger and thumb nail between the swollen part and the bit in the goat. Dress the wound with peroxide. If undetected the tick falls off when it has gorged itself, digests its meal and then crawls onto another victim.

Ticks spread diseases such as sheep scab, scrapie and other virus diseases. In areas where ticks are bad, goats should be routinely dipped in sheep dip.

POX

Symptoms Small red spots which rapidly develop into blisters similar to chicken pox, these burst and become infected ulcers similar to those in foot and mouth disease (see Infectious Diseases, page 126). The spots are usually confined to areas of bare skin such as udders. The goat may run a slight temperature at the onset of the disease.

Cause An acutely contagious virus, spread by external parasites, the goatkeeper's hands or contact with infected animals.

Treatment   The real trouble with goat pox starts when the ulcers have formed scabs which get broken during milking. *Do not* wash the udder with detergents or disinfectants, but apply carbolised vaseline or chlorhexidine ointment, which must be wiped off the teats with a disposable swab before milking. Reapply the ointment when milking is finished. Milk infected goats last. If this treatment is followed as soon as the trouble is noticed it should clear quite quickly. If, through neglect, secondary infections have entered, you should seek the help of your veterinary surgeon.

RINGWORM

Symptoms   Patches of thickened scaly skin can be seen where the hairs have fallen out. There does not appear to be any irritation.

Cause   Spores which live in dark buildings away from sunlight, therefore goats housed in dark, poorly ventilated buildings are the most likely victims. Such spores can live for several years in farm buildings and can also survive outside if sheltered from direct sunlight. They also infect humans.

Treatment   This is of little use unless housing conditions are improved. All wood must be scrubbed with a strong soda solution, and when dry given a coat of creosote. Improve the lighting and ventilation and clean out all old bedding and replace with clean straw. Medication for the goats should be obtained from the veterinary surgeon, since new and more effective remedies are continually being found.

SCURFY SKIN

This is a normal condition at the end of winter when the goats begin to cast their coats. On a warm day in spring wash them in a good cattle shampoo, lathering them well all over. Rinse and while still wet gently comb out the undercoat and loose hairs. After this they can be groomed regularly with a stiffish brush to tone up the skin and help encourage new growth.

Worm problems

This is the worst recurring problem facing every goatkeeper, and until a routine remedy is found and maintained his herd will never be 100 per cent productive. Worms can be just as much of a problem in housed goats as in those running on infected pasture, especially if there is condensation and the bedding is damp.

There are a number of different kinds of worms:

Nematode or round-worms which do *not* need an intermediate host to complete their life cycle. They migrate from the intestines to the lungs and cause parasitic bronchitis or husk.

Trematoda Flat leaf-like worms or flukes which need a specific species of snail found in damp meadows, and which flourishes in wet summers, to complete their life cycle. They infect the liver.

Cestoda or tapeworms. Their intermediate hosts are mites and fleas and they infest the intestines. They are sometimes expelled with dung as long segmented strips.

Coccidia which cause bloody scours. They are associated with wet drinking areas but need no intermediate host. (See Gastro-enteritis, page 116).

Treatment At one time it was necessary to use several different vermifuges to kill these different species, but so much research is going into this subject that better methods of control are continually being discovered. It is best to ask your veterinary surgeon to tell you what you should use, bearing in mind that it is advisable to use several vermifuges from different families of drugs during the course of a year, to ensure that the worms do not build up a resistance to any of them.

Prevention As well as treatment, some form of preventive control must be exercised in the goathouse and out on the pastures.

Worm eggs are voided in droppings and in moist damp conditions will hatch out in 10 days. They then crawl up the leaves of damp grasses ready to be eaten by the goat and so re-infect her. To combat this, move your goats to new pastures every week and either cut the grass or put horses or cattle on it to graze it down (see Management Tips in Chapter 1). Keep the bedding in the goathouses clean and dry.

All goats should be wormed 10 days after kidding and thereafter monthly until October, then twice more before starting the worming programme again in April. In-kid goats should be wormed 6 weeks before kidding.

Worm eggs can encapsulate themselves in periods of drought, and when the rains come and conditions are right they hatch out and the infestation in the pasture explodes to epidemic proportions.

12

SHOWS AND SHOWING

'Somebody's mother' hasn't, has she?

There are over eighty shows held each summer in the British Isles that are recognised by the British Goat Society, as well as innumerable unrecognised shows run by smaller agricultural societies and local goat clubs. Learner goatkeepers who wish to show their animals are advised to start their show careers at these smaller shows. You can learn so much by comparing your animals with those of the more experienced exhibitors. How do yours compare for size, conformation and condition? Exhibitors and stewards are friendly folk and, provided you ask your questions when they have time to answer, all will be well. Don't accost an exhibitor who is taking his goat into the ring or when he is struggling through the crowds with armfuls of fodder. Remember that you are asking for the advice of an experienced person who is going to give it to you free.

If you are interested in showing and want to have a go yourself, start off at a small club show. Those sleek coats and the beautiful condition you admire in the goats you have seen

148

winning are the result of good husbandry. The goats are healthy and well fed and this is the product of months of hard work. The bloom on their coats is reasonably easy to achieve if the groundwork has been properly done during the winter and spring. You can shampoo, groom, rug and do all the other little fiddles which you see experienced exhibitors doing when they are 'titivating' prior to going into the ring, but the result you get will depend entirely on how healthy your goat is.

As well as being properly managed through the winter, a goat needs training. We have found that, apart from bottle feeding, the simplest way of getting goats trained is to lead the kids out to pasture from a very early age. The lesson is short and they very quickly get used to the idea. They like going out to pasture so that leading is associated with pleasurable things. Try tying them to their pen gate for a few moments once they are used to being led. Then get them used to being tied and they will like it if you can give them something nice to eat at the end of it. In this way you get them biddable and they enjoy trotting along beside you. Then get them to stand still as you want them to when showing, head held alertly, feet in a natural and well balanced position. When you are in the ring the goats are lined up with their rumps towards the judge until they have been examined. They will then be made to walk round or across the ring and after this they are lined up sideways on to the judge.

To make the best of your animal you must keep an eye on the judge so that whenever he approaches your exhibit is looking at its best. Never have an earnest chat with your next exhibitor because, apart from taking your mind off what you are supposed to be doing, it is discourteous to the judge.

Preparing your animal for a show takes longer at the start of the season than it does after midsummer, by which time all the old winter coat will have been shed. About a month before the first show and when the weather is suitable, you should trim off the old beard and the long coarse hairs which sometimes grow between the front legs. Trim the feet. Shampoo well with a good cattle shampoo and rinse thoroughly. It is worth the extra expense of getting one of these because it makes the goat's coat look so nice. While the goat is still wet, use a strong comb, groom out all the undercoat. Rub the goat reasonably dry and put on a towelling rug. Leave this on overnight. If the coat is lying down and looks smooth there is no need to do more than groom it with a soft brush for a few minutes each day. If it is harsh and still curly you will have to rug the goat each night. About four days before the show shampoo again, trim the feet, and after this rug each night. This is quite sufficient to get the coat in good order. If this does not do the trick then the animal is not in fit condition and nothing you do outwardly will improve the situation.

Very nice-looking slip-on show leads can be made from plaited nylon cord. It is a good scheme to make some of these leads from baler cord in order to get practice. Leads made this way can be used for tying up and training, etc. Always cut your baler cord at the knots and then you have a useful piece of knot-free string.

Take three strands and fold in half to find the halfway mark. It is easier when you are learning to make these to get a second person to hold the strings while you plait, but you can manage with a tethered spring peg. Going back two inches from the halfway mark, clamp the three strings in the peg. Start plaiting and continue until you are two inches past the halfway mark. Unclamp the strings and bring both plait ends together and clamp. You should now have a plaited loop leading into the peg and six strands of string out of it.

Divide the six strings into three pairs and plait until the finished length just meets round the goat's neck. Peg here. Divide the strings into two sets of three and plait each set until they are eight inches long. Put a peg on each end. Now pass one of the strands of single plaiting through the loop you first made, straighten the two strands and make sure that they are the same length. Clamp the two ends and start double plaiting until you reach the very end of your string. Tie a single knot in the plaited string as close to this end as you can. Trim the ends.

You should finish up with a slip lead with a Turk's head type knot at the end of it. When you have got the hang of making the leads you can make your show leads with nylon chord. These leads are washable and look very smart, particularly white ones, and it is much easier to handle your goat and establish contact using this type of lead than it is with a stiff leather collar.

PATTERN FOR A GOAT RUG

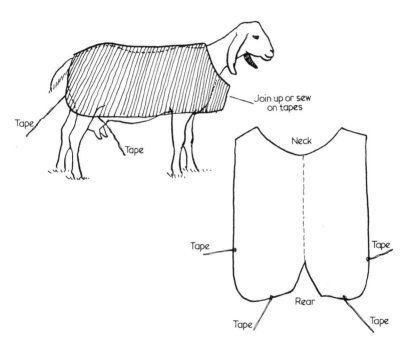

25 Pattern for a goat rug

Figure 25 shows you how to make a goat rug. If you put the tapes on as indicated you will not need a sursingle (girth), which often makes a wave in the hair just where you don't want it. Experiment with a bit of old material first. Measure your

151

goat from just in front of the withers to the hips, and from the hips to about an inch of the tail root; also from the withers to the front. The length of the material should reach from the front of the goat's chest to the tail and the width from the back seam to the top of the front leg. Cut along the top and tack up. Try it on the goat and get the right angle to cut for the neck and front seam join. Also cut along one side of the bottom and up to the tail root, as indicated in the sketch, while the material is still on the goat. Untack from tail to shoulder. You now have a pattern which will fit that goat.

To make a warm rug you can use an old blanket or coat and line it with an old sheet. Cut out two thick sides and machine from tail to shoulder. Cut out two thin linings and machine along the same seam. Putting the right side of both these seams together, tack or pin them along the seam. You now have the wrong side of the lining and the wrong side of the blanket on the outside. Put pins all over the two materials to make sure they are flat. Tack all round the edges and then machine all round, but leave nine inches unsewn so that you can turn the rug so that the right side comes outside.

Press all round the edge. If you want to be very smart, at this stage you tack on coloured braid round the edges. Several rows of machining right round the rug will make the edges smart enough and strong enough.

Try the coat on the goat again and you will be able to decide where to sew up the front and how long to make your tapes.

These rugs can be made in cotton for the summer when it is warm, or in towelling for use after bathing. It is much better to line the thick ones with sheeting as this makes it easier to see how dirty they are getting.

There is nothing more damaging to goatkeeping than when goatkeepers exhibit their animals at agricultural shows with dirty rugs or bits of material tied up with string. There is no excuse for it either. If you take your animals in the show ring unwashed, or stained from wintering on dirty beds, or with tatty collars, you cannot expect to be judged properly. It takes a lot of time and energy to get to a show and it is just plain stupid to spoil your chances in this way. Scores of goatkeepers do it too. It is no good going home grumbling that Mr So-and-so always wins. Mostly he wins because he takes the trouble to train his animals properly and to turn them out as smartly as he can. Your goat may be as good as his, but nobody will notice,if it is unkempt and will not behave itself in the ring. A show like this is the goatkeeper's shop window and only the best is good enough.

All recognised shows print schedules and entry forms which must be filled in and returned to the office by a certain date, together with the entry fee. There are always moans and grumbles about the length of time between the closing date for the entries and the actual show, but I can assure you that shows keep this time down as much as they can, but are bound by the time it takes to check the entries and prepare the catalogues for the printers; and then the printers need time too.

Having prepared your goat you will also need two buckets and a milking bucket, a hay rack or net, and enough concentrates, hay and green food to last her through the show. The latter must be the same as she is getting at home, and it usually takes up a lot of room. She needs her numbers, show leads and a collar and lead to tie her in her pen when necessary, and grooming kit and rubbers, for she is sure to get some dirt on herself at the last minute. Don't get into the habit of bathing your goat the night you arrive at the show. It may be too cold or too late and in any case it makes it difficult for her to settle down quickly. For yourself, if you are staying overnight – and you have to at all shows which run milking trials – you need a camp bed and a good sleeping-bag and blanket, and a picnic stove on which to boil a kettle and cook your breakfast. Your food, pans, cup and plate and cutlery, a clean white overall, rubber boots and mac . . . the list can get endless. If you are going to show regularly, make a list on a piece of card and add to it all the things you forgot. It is a valuable piece of equipment!

Get a catalogue. Study the order of judging and be sure you are ready when your class goes in. The novice exhibitor nearly always falls into the trap of thinking that is the end when he comes out. It is not. There is the Inspection-Production line-out and all milkers placed down to the reserve *have* to parade for this. It takes place after the AOV milkers' class. Then study the special prizes. Many of these are for novice exhibitors and you should find out from the steward when these are to be judged and make sure you are ready when he calls for you. This is particularly important at day shows and club shows where cups and special prizes are always given for novices.

Showing can be a lot of fun. Be serious, but not too serious because it is not a matter of life or death if you lose. I can't think of anyone in this country whose livelihood entirely depends on always winning. To those who scorn showing, could I point out that unless we had some other completely different breeding system, any improvement in goats would be difficult to encourage without this kind of publicised competition.

APPENDIX A
THE BRITISH GOAT SOCIETY

The British Goat Society was formed in 1879 with the object of circulating knowledge and general information about goats and 'with a view to counteracting the prejudice and ignorance which prevail in a great degree concerning these animals' (Holmes Pegler). The founders of the society also wanted to encourage 'cottagers' so that milk could be produced in rural areas where it was frequently unobtainable, and to improve and develop those qualities in the goat which were recognised and valued in cows.

Over the years the Society has grown and so have its undertakings. It now registers and records all goats in this country which belong to members of the Society or to members of Affiliated Societies. It also records all transfers, so that members, by searching the *Herd Books*, can trace the pedigrees of all the goats that the BGS has ever registered.

The Milk Marketing Board organises for the Society a scheme for weighing milk and sampling it for butterfats, which covers a period of 365 days from the fourth day after kidding, with a further extended lactation if desired. At one time the Society ran the Stud Goat Scheme, with the financial assistance of the Ministry of Agriculture and Fisheries. Approved males were inspected before being accepted for this scheme, and stud owners were expected to provide a service for cottagers and smallholders and those in similar circumstances. Fees were limited to five shillings, and the Ministry paid a further five shillings for each service at the end of the season. This scheme was abandoned some years ago, but it did a great deal to improve goats all over the country. Many a good-looking 'scrub' goat seen today is a descendant of a Stud Goat Scheme male.

The British Goat Society also 'recognises' certain shows. These shows have to be judged by a person licensed by the BGS as a judge. The society has to approve all the classes in the 'recognised' section of the schedule and any milking trials which take place have to be run strictly according to BGS rules. At these shows goats can win open and breed challenge certificates and can win special awards for the quantity of milk produced and the butterfats they yield. By winning a series of challenge certificates and breed challenge certificates, goats can become champions or breed champions. Since these certificates depend on a minimum amount of milk given in the trials it follows that a champion is a very worthy goat indeed.

Milk-recorded goats are given decorations according to their yields; *R20* for one which has given 2,000lbs in twelve months, *R36* for 3,600lb and so on. A few goats have even got as high as *R65*! If the daughter of an *R25* goat yielded 3,500lb she would go into the register of merit and become *RM35*. The highest step up this ladder is the Advanced Register or *AR* prefix. It is not my intention to give you all the details of either the prefixes awarded for milk records or those prefixes or affixes which can be won at shows, because these can be altered from time to time. If readers want more information on the subject they should send a stamped addressed envelope to the British Goat Society asking for the latest regulations.

The BGS is becoming increasingly concerned by the number of goats who lose their identity and registration numbers because owners either lose the registration card or fail to transfer when the goat is sold, or die suddenly leaving a herd of goats, sometimes famous ones, which nobody else can really identify.

The Society has a scheme for earmarking goats. Individual breeders can register an earmark letter code, or Affiliated clubs can register one. A book of earmarking certificates is issued with counterfoils. When an animal is earmarked the certificate is sent with the registration card to the BGS. If everyone who earmarks a goat supports this scheme no two tattoos will be the same and this should make it possible to give a positive identity to any goat bearing one of these ear numbers. At present it is compulsory to earmark all kids before you can register them. You should contact your local goat club for further details. There have been countless occasions when my partner and I have come across a goat whose identity we almost certainly knew, but without proof we could do nothing. In cases like this of a goat which has had a breed section registration number which has been lost, all the unfortunate owner can do is to register it in the Identification Register and start on the long road back into the register to which the goat belongs. It is a long road too, because even if you are lucky it takes five *generations*, but mostly it takes seven or longer.

Many smaller clubs and societies are affiliated to the BGS. In the 1975 *BGS Year Book* fifty-eight are listed. Of these eleven are agricultural or show societies who are affiliated because of their goat section. Two are Australian and New Zealand goat Breeders associations and the rest are county or district goat clubs scattered around the British Isles. It is these clubs which newcomers to goatkeeping should join first, because they can join in with all the local goatkeeping activities and meet people who will be able to help and advise them with all their problems. They will be able to register their stock through the clubs and then later, when they become more seriously interested in goatbreeding, they should join the parent society. Clubs have much cheaper membership fees, but registrations cost more. On balance it is best to join a club first in order to learn the ropes and then if you do not intend carrying on you will not have spent too much money! The Secretary of the BGS will supply you with a list of Affiliated Clubs if you send a stamped addressed envelope for the reply.

I find it a sad reflection that Holmes Pegler's hope of counteracting prejudice and ignorance about goats has made so little progress in spite of valiant efforts on the part of certain past and present goatkeepers. It is interesting to note, too, that in 1975 fresh or pasteurised cows' milk is just as difficult and expensive to obtain in some rural areas as it was a century ago.

Many goatkeepers want an artificial insemination service in this country. In France where groups of goatkeepers are situated round certain cheese factories it is comparatively simple to run a viable artificial insemination service, but goats are scattered pretty widely and thinly round the British Isles and this in itself is a formidable problem. There is an effective scheme running in Canada and in the United States and conception rates in all three countries are fairly satisfactory. The BGS have regulations which cover the registration of kids born as a result of artificial insemination and no doubt further progress will be made in this direction when enough goatkeepers give active help and co-operation to the scheme and, more important, when enough herds want to use the scheme to make it viable.

Code of Practice for Goats

This code has been drawn up by the British Goat Society as a leaflet. It reads:

This leaflet is not a treatise on how to keep goats properly, but a list of that animal's basic needs and is for the use of Officials when investigating cases of neglect etc. Animals subjected to conditions poorer than listed below can be said to be suffering from neglect in some degree according to the case in question.

HOUSING

A goat should have access to adequate shelter at all times. It should be housed at night all the year round and also in bad weather. These animals suffer acutely in wind, incessant rain, flies and hot sunshine.

The house should have a suitably situated window, adequate ventilation but free from draughts. There must be a hay rack high enough for the goat to reach in comfort but not less than three feet so that the hay will not be soiled by the goat. There must be a daily supply of fresh, clean water. These animals will starve to death rather than eat food trampled under foot and they will become dehydrated rather than drink fouled water.

Each animal should have at least four by four feet of floor space and horned and hornless goats must *Never* be penned together. Since goats are very active animals they should never be tied when housed.

There must be adequate dry bedding of either straw, shavings or other suitable material.

FEEDING

Providing that FRESH, CLEAN pasture and browsings are available a goat can maintain itself without discomfort from May until August. After that palatable hay must be added to the diet and later an adequate supply of fresh foods, roots, kale, vegetable trimmings and other clean household scraps etc.

If housed, a daily ration of 4 to 5lb of hay and the equivalent of 10lb kale must be fed daily. Ideally some concentrates should be fed in winter, crushed oats being the most suitable of the cheaper foods.

Mineral licks must be available at all times.

Kids must have milk for at least 4 months and longer if they are not fed supplementary concentrates. Most calf milk substitutes are suitable for kids.

CONTROL

The worst method of controlling goats is tethering since this is seldom done adequately. A light chain not less than 10 feet long with at least two swivels and a light strong leather collar must be used. This must be moved at least twice daily. The goat must have access to shelter when tethered if the owner cannot move it when the weather turns, and it must have a supply of clean water too. If tethering is practised during the day the goat must be free in its house during the night. It is most important that the goat should be tethered on suitable grazing, being moved twice daily. It is not advisable to tether goats when browsing due to danger from entanglement.

GENERAL

1   When in milk the goat should either suckle a kid or have the milk removed from the udder at least once daily and twice for the first six weeks after kidding.

2   The feet must be adequately trimmed.

3   It is recommended that a dressing of louse powder be applied if necessary.

4   Goats should be dosed regularly for worms with a suitable preparation as recommended for sheep.

5   If a goat is sick or injured the law requires that it shall have skilled attention.

6   The sale of kids in the open market is not recommended.

7   Kids should never be tethered.

8   If a person cannot or will not provide at least the above maintenance for their goats they should not keep such animals.

APPENDIX B
USEFUL ADDRESSES

ADAS   This advisory service is listed in your telephone book under Agriculture, Fisheries and Food, Ministry of . . . Division.

American Dairy Goat Association   PO Box 186, Spindale, North Carolina 28160, USA.

British Goat Society   Rougham, Bury St Edmunds, Suffolk, IP30 9LJ.

Express Rubber Stamp Services Ltd   Channing Works, 94 Mill Road, Kettering, Northants, NNL6 0TP.

Hansen's (CHR) Laboratory Ltd   476 Basingstoke Road, Reading, RG2 0QL. (Booklet on cultures, Annatto cheese and butter colouring, cultures for yoghourt, lactic ferment cheese and butter starter, and cheese rennet.)

Lakeland Plastics Ltd   Alexandra Buildings, Station Precinct, Windermere, Cumbria.

APPENDIX C

GOATS' MILK IN INFANT FEEDING

(condensed from a paper by J. B. Tracey, MB)

'What exactly do you have to do to prepare feeds when you put an infant on to goats' milk?' Every Goatkeeper must sooner or later be asked this question; and I propose to give you a simple answer which I have given in the many cases that I have put on to goats' milk for medical reasons, and which has resulted in the rearing of so many successful, trouble-free, healthy infants, that I believe that more complicated instructions are unnecessary.

First let me emphasize that I am assuming that the hygiene of production of the milk is satisfactory, so that one may safely recommend that the milk can be given raw—that is without heat treatment by boiling or pasteurization. I do not believe that boiled goats' milk is ever quite the same as the fresh raw product. I think that the curds are probably of different physical properties; that the fat is apt to separate from the curds; and that the lactalbumin is coagulated or solidified to form a skin which may delay the rapid digestion which is quite the most important advantage of Goats' Milk.

The principal reason for the almost universal recommendation by Doctors and Nurses to boil all milk that is to be fed to babies, is the fear that the milk may be infected with tuberculosis. With goats' milk I am confident that there is *No* Danger of tuberculosis infection in milk from ordinarily healthy Goats.

Goats' milk is so readily digested by infants that I have found in a number of cases that, where the breast milk supply has been found to be failing to provide enough to satisfy the baby, full strength goats' milk can be given preferably from a spoon, to make up the necessary quantity to satisfy the infant, while, unsweetened, this will not tempt the infant to wean itself on to the bottle. I have found that infants of mothers who have been going to work have thrived on a midday feed of goats' milk, welcoming the return to mother's milk at the other feeds.

Appendix C

Where the change on to goats' milk is being made owing to digestive upset when on a cows' milk preparation, I recommend that the first few feeds be made up at half strength (half boiled water/half goats' milk), and, when all the clots of cows' milk have been passed downwards or vomited, the strength should be increased in a few feeds to two-thirds and then three-quarters, reaching full strength in, at the most, two or three days. A suggested quantity per day would be from 2 to 2¼ fluid ounces per lb body weight, the smaller infant having the greater relative amount.

Sweetening with granulated sugar may be needed and acceptable when weaning, as breast milk is very sweet to the taste and the infant would miss this. The quantity can readily be judged by taste, one or two level teaspoonsful being tried in each feed at first. Honey or demerara sugar can be used if preferred.

It is possible for babies fed either on the breast or weaned from the breast on to goats' milk only, to develop an unusual type of anaemia due to a shortage of iron and of folic acid in both human and goats' milk. This is easily overcome by getting your doctor to prescribe these in addition to the usual vitamin preparations given to infants at this stage of life. When weaning starts on to mixed food, these defects will easily be overcome in the usual little messes of broths and purées which are added to the diet.

Repeatedly have I found that infants which have failed to thrive for no apparent reason, have turned the corner on goats' milk, and I have only heard later that the child was kept on it for a length of time, eventually weaning itself from all milk feeds when in full and normal health. To summarize: I believe that fresh, raw, hygienically produced, undiluted, slightly sweetened, and fed at blood heat, goats' milk will overcome most digestive upsets, and rear healthy, strong infants to weaning stage and after, provided that the usual vitamin supplements, etc, are given.

APPENDIX D

GOATS' MILK IN TREATING INFANTILE ECZEMA

(condensed from a paper by J.B. Tracey, MB)

Infantile Eczema is an inflamed condition of the whole or part of the skin *starting in the first year of life* and extending into the second year, and tending to cure itself by the age of two. The condition passes through three phases: Redness of the skin with itching and scratching; Breakdown of the skin with weeping and scab formation; Scaling or flaking off of the scabs, with further itching. *Infection* may occur at any phase due to scratching.

In the *first and second stages* it is usual to apply a simple lotion or powder prescribed by the doctor, who will also probably give a sedative to lessen the infant's frenzied desire to scratch, and this will be continued into the *third stage* when some oily or greasy preparation will be used to help the scales to separate.

It has been noticed that, in a large proportion of cases of infantile eczema, the first onset of the condition follows shortly after the beginning of weaning from the breast on the cow's milk or one of its derivatives, or shortly after birth if never on the breast. Many such cases improve on goat's milk. It is well known that many infants, unable to digest cows' milk or its products satisfactorily, do well when changed on to Goats' Milk. It has also been found that some cases of allergy (asthma or hay fever) improve on goats' milk. It is believed by some that the cause of infantile eczema in some cases, if not all, may be a reaction of the skin to the proteins in cows' milk taken in feeds, so that, when this is entirely replaced by goats' milk in the diet, there is a rapid improvement and eventual cure of the condition, as long as no cows' milk or beef proteins in any form are given up to the age of two years.

THE EARLY MONTHS

Eczema in an entirely breast fed baby is a comparative rarity, and it

163

Appendix D

may be assumed that at least partial weaning, complementary or supplementary feeding with cows' milk or one of its many derivatives has started before the onset of the rash. If so, all cows' milk, etc., should be stopped and replaced by goats' milk which can be warmed as usual and given in most cases undiluted.

Where the baby is either upset in the stomach or actually vomiting cows' milk or other feeds, it may be wisest to start with small quantities of goats' milk at half strength for the first few feeds only, gradually decreasing the boiled water added to dilute the goats' milk until, within two or three days at the most, full strength is given. Some of the first feed or so may be vomited in these circumstances until all the unsatisfactory feed which is being replaced has been eliminated, so this should not be a cause for worry. The infant can safely be allowed as much of the goats' milk as it will take, for the start, at the intervals to which it is used. If in doubt, consult your doctor or nurse.

WEANING

Which may be defined as the addition of food to the diet to replace an increasing proportion of the breast milk, may start at any age from birth. It is most common to introduce a large proportion of cows' milk or one of its dried products at the start, and gradually introduce cereals, vegetables and meat dishes. As soon as infantile eczema makes it appearance, it has been found best to stop all additions to the breast feeds and give only goats' milk in their place until some definite improvement occurs, and it is at this stage that the utmost care must be taken to ensure that no beef proteins are included in any of the items introduced into the weaning diet.

NEVER give any of the patent foods commonly used without first studying the analysis on the carton, etc. Reject any that contain unwanted beef protein. If in doubt ask your chemist to give you his opinion in the light of this article which you can show him.

It may be wisest to restrict the carbohydrate or starchy additions to the diet to *plain ingredients* such as flour, cornflour, semolina, rice, oatmeal, wholemeal flour, etc, made up yourself with potatoes and vegetables sieved as usual and flavoured.

The goats' milk will supply all the protein needed in the diet and a large proportion of the fat, though additional goats' cream and goats' butter may be an advantage, if obtainable. Much of the goats' milk will be given in cooked form: milk puddings, custards (when eggs are admissable), vegetable soups, etc.

Margarine is a useful ingredient of the diet, and contains added Vitamins A and D by law, so long as no cows' butter has been added.

Opinions are divided on the role of sugar in the diet but, in the theories under discussion it should be harmless. That admits Sugar, Treacle, Honey, Jam and Marmalade; and fruit and fruit juices can be sweeted and will help to provide Vitamin C.

When the eczema has subsided great care must be taken in adding meat proteins to the diet, as one or other of these are believed to be the essential cause of the eczema. It is wisest to introduce a single class of meat at a time, and continue its use for a period of at least ten days or a fortnight, as the only new addition over and above the diet on which the rash has subsided, as it is commonly found that it may take this period of time for sensitivity or allergy to show itself. As soon as this period of time has passed without rash recurring, a further class of meat or protein can be added. Should a rash or reaction follow the new item, this should immediately be dropped from the diet and not be tried again until cure has been complete and after the second birthday at earliest.

It has been reported in the early stages of the use of the goats milk in many cases that there is a temporary flare up of the eczema, with weeping and reddening of the skin, and this has rapidly been followed by the first real improvement. At the same time, however, the general condition and improvement in sleeping that has been a hopeful and instant response to the new diet, is often maintained throughout this apparent setback. A growing weight of evidence thus exists for the value of goats' milk in replacing cows' milk in the diet of infants suffering from Eczema, the whole treatment being based on the theory that goat's milk provides an acceptable substitute, and does not in itself produce a cure when used in addition to cows' milk.

APPENDIX E
PHYSIOLOGICAL DATA FOR THE GOAT

(from *Management and Diseases of Dairy Goats*)
by Samuel B. Guss

Parameter
Temperature (rectal): 101.5°F–104.0°F
Pulse: 70–80/min.
Respiration: 12–20/min.
Rumination: 1.0–1.5/min.
Puberty: 4–12 mnths.
Oestrum (Length of Heat): 12–36hrs. (Avg. about 18hrs.)
Oestrous Cycle: 18–23 days
Gestation: 148–153 days (Avg. 150)
Birth Weight: Dependent upon breed and number (from 4.5 to 9.0lb., or 2.2 to 4.1kgms.)
Urine Specific Gravity: 1.015–1.035

Haematology
Blood Volume: 7.0% body weight
Hæmoglobin: 8–14
Packed Cell Volume (PCV): 30–40%
Haematocrit/Red Blood Cells (RBC): 8–17.5 millions/ml.
Haematocrit/White Blood Cells (WBC): 6–16.0 millions/ml.
Clotting Time: 2.5 minutes

Differential

Neutrophils: 30–48%   Monocytes: 1–4%
Lymphocytes: 50–70%   Eosinophils: 3–8%
Basophils: 0–2%

Clinical Chemistry
Blood Glucose: 40–60mg%
Blood Urea Nitrogen: 13–28mg%
Creatinine: 0.9–1.8mg%
S.G.O.T: 50–100 S.F. Units
S.G.P.T: 12.17 S.F. Units
Calcium: 9.5–10.5mg%
Magnesium: 2.5–3.5mg%
Phosphorus: 3.8–7.6mg%
Sodium: 135–154mg%
Potassium: 3.9–6.3mg%
Chloride: 105–120mg%
Ca: P Ratio: 1.7:1 in milking does; 2.0:1 in others

INDEX

Page numbers in italic represent illustrations

Index